FRUIT PRODUCTION

NIPA GENX ELECTRONIC RESOURCES & SOLUTIONS P. LTD.
New Delhi-110 034

FRUIT PRODUCTION
(Minor Fruits)

Rakesh Kumar
Assistant Professor
Fruit Science
Rainfed Research Sub-station for Sub-tropical Fruits, Raya
Sher-e-Kashmir University of Agricultural Sciences and Technology of Jammu
Jammu-180 009, Jammu and Kashmir

Vijay Kumar
Assistant Professor
Soil Science and Agricultural Chemistry
Rainfed Research Sub-station for Sub-tropical Fruits, Raya
Sher-e-Kashmir University of Agricultural Sciences and Technology of Jammu
Jammu-180 009, Jammu and Kashmir

Reena
Associate Professor
Entomology
Advance Centre for Rainfed Agriculture, Dhiansar
Sher-e-Kashmir University of Agricultural Sciences and Technology of Jammu
Jammu-180 009, Jammu and Kashmir

Parshant Bakshi
Associate Professor
Division of Fruit Science, Chatha
Sher-e-Kashmir University of Agricultural Sciences and Technology of Jammu
Jammu-180 009, Jammu and Kashmir

V.B. Singh
Associate Professor
Plant Pathology
Division of Fruit Science, Chatha
Sher-e-Kashmir University of Agricultural Sciences and Technology of Jammu
Jammu-180 009, Jammu and Kashmir

NIPA GENX ELECTRONIC RESOURCES & SOLUTIONS P. LTD.
New Delhi-110 034

NIPA GENX ELECTRONIC
RESOURCES & SOLUTIONS P. LTD.

101,103, Vikas Surya Plaza, CU Block
L.S.C.Market, Pitam Pura, New Delhi-110 034
Ph : +91 11 27341616, 27341717, 27341718
E-mail:newindiapublishingagency@gmail.com
www: www.nipabooks.com

For customer assistance, please contact
Phone: + 91-11-27 34 17 17 Fax: + 91-11- 27 34 16 16
E-Mail: feedbacks@nipabooks.com

ISBN: 978-9389130-17-1

Composed and Designed by NIPA.

Preface

Agriculture is an ancient practice which is being performed by the human since prehistoric times. In India agriculture is one of the oldest occupations. Among the agricultural branches, horticultural is one of most important and high pay back field in arid, semi arid, degraded and poor soil condition areas. Fruit science plays significant role in the social economic life of the people, are directly or indirectly depend upon our basic requirement of nutrition in humans. The scenario of horticulture crops in India has become very encouraging, during 2016-17, the production of horticulture crops was about 295.2 million tonnes from an area of 24.9 million hectares. Significant progress has been made in area expansion resulting in higher production. Over the last decade, the area under horticulture grew by about 3% per annum and annual production increased by 5.4%. (Horticultural Statistics at a Glance, Govt. of India, 2017). In fact, the horticulture has also gained commercial importance with a very significant share in the economy of the region. Fruit crops *viz*., mango, apple, pear litchi, guava and citrus are now considered to be the most ideal strategy to provide nutrition and income security to the fruit growers in all climatic zones. Moreover, now these days the importance of minor fruit crops *viz*., Bael (*Aegle marmelos*), Phalsa (*Grewia asiatica*), Jamun (*Syzygium cuminii*), Aonla (*Emblica officinalis*), Ber *(Ziziphus mauritiana)*, Fig (*Ficus carica*), Tamarind (*Tamarindus indica*), Galgal (*Citrus pseudolimon*), Lime *(Citrus aurantifolia),* Karonda (*Carissa carandas*) and Jackfruit (*Artocarpus heterophyllus*) are very important in improving the productivity of the degraded land, rainfed areas, drought areas and improving social economic conditions of the small and marginal fruit growers. Minor fruit crops have assumed significant importance in the crop diversification, which has become essential to arrest serious land degradation and enhancing the farmer's income. All these fruits are the rich source of diverse health building substances especially protein, vitamins and minerals and offer an advantage for food and nutritional security. Besides, their high yield potential per unit area and with varied agro-climatic conditions ranging from tropical to sub-tropical in India. Moreover, there fruit crops are serving in those areas where another crops *viz.,* cereal, pluses and vegetables are quite difficult to cultivate due to lack of irrigation facility and low rainfall condition. It is creating diverse varietal wealth, labour intensiveness, high market price and prospects of processing, value addition and export they play an important role in employment generation and livelihood in rural and far flag areas. Fruits *viz.*, jamun, karonda and phalsa are highly perishable, jackfruit and custard apple contains high sugar content while bael fruit is not an easy to eat out of hand item. Many people because of its strong astringent taste do not like fresh aonla fruit. But all these fruit can be utilized in processing products with the help of some value addition. Manual of minor fruits will provide information on different aspects of fruit production *viz.*, fruit crops name, identification of field tools, application of fertilizers, training and pruning different inputs used in raising fruit crops including tools and implements and procedures involved in nursery raising. It also impart knowledge to the students. The students have been provided with exercises and data sheets to have a practical insight of the subject. Production of quality planting materials from both sexual and an asexual methods in rainfed areas may be provided for the students to inculcate practical skills and better understanding of the subject. This practical manual and theory of minor fruit is an effort to sensitize the students horticulturists, research scholars and agriculture extension officers about practical aspects of minor's fruit production and will motivate them to choose these fruit crops to increase the income of fruit growers.

Authors

Contents

Preface .... *v*
Glossary .... *ix*

1. Classification of Fruits 1
1. Fruit : Types and their Classification 1
2. Angiosperms 4
3. Based on Rate of Respiration 4
4. Based on Light Requirement 4
5. Based on Water Requirement 11
6. Based on Photo Periodic Requirement 11
7. Based on Branching Habit 11
8. Based on Longevity 11
10. Inforescence 11
11. Based on Type of Pollination 12
12. Leaf Morphology 14
13. Horticultural Zones of India and their Classification as Plant Behavior 16
14. National Fruits of Different Countries 21
15. Based on Rate of Ethylene Production 21
16. Based on Storage Life 21

2. Use of Field Tools in Fruit Nursery 25

3. Nutrient Management in Fruit Crops 29
Soil 29
Objectives 29
Organic manures 30
Farmyard manure (FYM) 30
Vermicompost 31
Green manures 31
Sources of manures 31
Bio-fertilizers 32
Benefit of human health 32
Benefits of organic manures 33
Inorganic fertilizers 33
Compound fertilizers 33

Identification and nutrient content of chemical fertilizers 34
Procedure/methodology 34
Calculation of quantity of fertilizers based on nutrient requirement 35
Calculation of quantity of complex fertilizers 35
Calculation of quantity of chemicals for spray solution 35
Methods of fertilizer application in fruit crops 35
Methods of fertilizer application 35
Criteria for land capability class 38

4. Cultivation of Minor Fruits 43
Importance of minor fruits 44
Necessity 44
Planting, manures and Fertilizers 47
Cultivars 51
Propagation methods 52
Harvesting and yield 55
Insects, Pests and Diseases 57

5. Layout and Planting of Orchard 73
Layout plan 73
Aims of layout plan 73
Types of layout 74
A. Square System 74
B. Rectangular system 74
C. Quincunx system 75
D. Hexagonal/Triangular system 76
E. Contour system 77
Layout planning in minor fruit plants 77
Planting distance of major and minor fruit plants 77

6. Plant Propagation and Its Methods 79
Plant Propagation 79
Germination of seed 79
Advantages of sexual propagation 79
Disadvantages of sexual propagation 80
Vegetative propagation 80
Grafting tools 83

7. Orchard Management under Rainfed Condition 95
Selection of orchard soil in rainfed areas 95
Important factors of orchard management 95
Soil management of fruit orchard in rainfed areas 96
Cultivation of cover crops 97
Advantages 97

8. Training and Pruning 99
Importance of training and pruning 99
Training 99
Objectives 100
Details of training 100
Types of training system 100
Overhead trellis or Telephone system 102
Tatura trellis 102
Pruning 102
Make pruning cuts appropriately 103
Purposes of Pruning 106
Crossed branches are unhealthy 106
Dead wood 106
Thinning cuts 106
Heading cuts 106
Dormant pruning 106

9. Research Organizations in Horticulture 107

Glossary

Abscission	Separation of an organ from a plant
Acre	43,560 ft^2 a unit of land
Acid	Compound that gives off H^+ ions in solution
Acidic	Describes a solution with a high concentration of H^+ ions
Adventive embryony	This is also known as nuclear embryony or polyembryony. In this case more than one embryo develops in a single seed. In the seed both types of embryo develops i.e. nuclear embryo from nuclear cell and zygotic embryo from egg cell with the result of syngamy.
Androecium	Collectively stamens of a flower
Anther	The swollen, apical, pollen-bearing section of the stamen
Anthesis –	Time of flower expansion when pollination takes place
Anthocyanin	A class of water-soluble pigments responsible for the red, purple, and blue coloration of flowers and fruits
Areas	Measures the size of a surface using length measurements in two dimensions
Atmospheres	Common units for measuring pressure
Androgamy	Development of the embryo from male gametes inside or outside the embryo sac is known as androgamy. Since the cells are haploid in nature they come under non recurrent type
Ascorbic acid	Vitamin 'C'
Apogamy	Development of embryo from synergids or antipodal cells within the embryo sac with or without pollination but without fertilization is termed as apogamy. This type of apomixis is also grouped into haploid and diploid apogamy depending upon the ploidy level of cell. Diploid apogamy is recurrent type whereas, haploid aopgamy is non-recurrent type.
Apomixis,	(Syn agamospermy) – reproduction of a plant through a seed wherein the embryo has arisen clonally from nucellar or integument tissue, and is genetically identical to the parent plant
Astringency	A drying sensation in the mouth

Auxin	Plant growth regulators used into root cuttings formation, chemically thin fruit, and prevent pre-harvest drop of fruit crops
Asexual	Without sex, as in vegetative reproduction or propagation
Axil	The angle formed at the point of insertion of a leaf to a stem
Axillary bud	A bud found in a leaf axil
Ayurvedic	Preparations ayurvedic treatments are primarily dietary and herbal. Ayurvedic medicine claims to be the traditional medicine of India
Base	Substance which gives off hydroxide ions (OH^-) in solution
Basic	Having the characteristics of a base
Dormancy	A temporary suspension of visible growth in organs containing meristems; occurs each winter in temperate fruits *viz.*, apple, pear, persimmon and plum etc.,
Buffer solutions	Solutions that resist changes in their pH, even when small amounts of acid or base are added
Baume	A scale for specific gravity, used to determine sugar syrup concentrations
Blanching	Heated rapidly in boiling water or steam, held for a known time and then
Cooled	Cooled quickly to near room temperature. Blanching is used to destroyenzymes and some micro-organisms
Blender	A mixer with whirling blades that mixes, chops or liquefies foods
Bisexual	(syns hermaphroditic, perfect) – both sexes are occurring on same plant
Brine solution	A strong solution of salt and water used for pickling
Brix	Measurement of the sugar content
Central leader	The central trunk in a central leader pruning system from which the main scaffold limbs emerge
Crotch angle	The angle a scaffold limb makes with the main trunk or a shoot makes with a scaffold limb
Calcium chloride	A soluble salt
Carbon dioxide (CO_2)	A colourless, odourless, non-poisonous gas that results from fossil fuel
Combustion	Cash flow the cash remaining in a month when all cash inflows are subtracted by all cash outflows
Caustic soda	Common name for sodium hydroxide, e.g. used for unblocking drains
CAZRI	Central Arid Zone Research Institute at Jodhpur, India.
Chimeras	A chimera is an individual with one genotype in some of its parts and another genotype in the others. Somatic mutation may often lead to chimeras. When propagated asexually these chimeras may become perpetual. Certain types of *Pelargoniums* and potatoes are of such chimeras. When

	growth is encouraged from the concealed tissues the real nature of these chimeras is revealed.
Cm	(= centimetre) unit of length in a metric system, equal to 10 millimetres
Cultivar	A race or variety of a plant that has been selected intentionally and maintained through cultivation
Citric acid	An organic acid which is naturally occurring in citrus fruits and used to adjust the pH of products
Dry Land Horticulture	Plant Propagation and crop production in arid and semi arid region with annual rainfall less than 600mm
Dehydrating	Removing water from a product
Density	The mass of a substance in a unit volume
	Dry ice solidified carbon dioxide
Dichogamy	In hermaphrodite plants, stamens and pistils mature at different times orreceptivity does not coincide with pollen viability/maturity and prevents self-pollination inperfect as well as monoecious flowers.
Ethrel	A chemical (ethephon) which is used as a plant growth regulator and for the removal of the astringency of persimmon fruits. Ethyl alcohol same as ethanol or alcohol
Ethylene	Gas a plant growth-regulating gas, naturally produced in ripening fruits, responsible for promoting ripening
Fermentation	Is a process whereby beneficial bacteria are encouraged to grow. These bacteria increase the acidity or alcohol content of a food and therefore prevent the growth of spoilage and food poisoning bacteria
Floriculture	It is branch of Horticulture which deals with cultivation of flowers and ornamental crops
Fruit breeding	Fruit breeding is the manipulation of a biological system that requires many generations to achieve result. It is also a dynamic, exciting and challenging profession, operating under continually changing conditions
Fumigate treat with fumes	With the aim of disinfecting or eradicating pests
Fungicide	A substance or chemical that kills fungi
g (= gram)	Basic unit of mass in the metric system, 1,000 grams equal 1 kilogram
	Gauge a standard measure used to determine the thickness of films
General or Common Nursery	Plant Nursery where all type of seedlings and plants are produced on a large and small scale
Genetic incompatibility	Means that a plant cannot produce a zygote with its own pollen.There are two types, sporophytic and gametophytic. If the pollen is of the same allele as that of the stigma, then pollination will not be successful.

Growth regulator	A natural or chemical substance that changes the shape and appearance of the plant, stops or starts plant growth, prevents flowering and other functions
Hi-tech Nursery	Fully automated nursery in which all environmental factors are controlled and manipulated for production of healthy and quality planting materials
Horticulture	Branch of agriculture which deals with cultivation of fruit and plantation crops, vegetables and ornamentals plants.
Hygrometer	An instrument used for measuring relative humidity
Impotence from abortive flowers	In perfect flowers, reflexed stamens partially defective pollen, or quantity of pollen produced is small, non-viable (e.g. European varieties of Grape) resulting in failure to set and mature fruit. Abortive flowers are more common in plantshaving indeterminate inflorescence. High pollen viability is necessary to ensure good fruit setand yield.
Impotence from degenerated or abortive pistil or ovule	Any factor that hinders theprocess of fertilization results in unfruitfulness or sterility. Complete: no flower or no sexorgans are formed or fail to attain full development; Partial: either stamens or pistils abortive,sometimes looking normal; Abortion of ovules: *e.g.* mango; multiple ovules and anthers resultin both female and male sterility.
kg (=kilogram)	Unit of mass in a metric system, equal to 1,000 grams
Loan	Capital in terms of money from financing agencies and institutions like banks and government schemes
m (=metre)	Unit of length in a metric system, equal to 100 centimetres or 1,000 millimetres
Male sterility	Male sterility is characterized by non-functional pollen grains, while female gametes are functional. Male sterility can be classified into three groups viz., genetic male sterility, cytoplasmic male sterility, and cytoplasmic genetic male sterility.
Mechanization	Nursery operations carried out with the help of machines
mg (=milligram)	Unit of mass in a metric system. 1,000 milligrams equal 1 gram
ml (=millilitre)	Unit of volume in a metric system. 1,000 millilitres equal 1 litre
mm (=millimetre)	Unit of length in a metric system. 1,000 millimetres equal 1 metre
L (=litre)	Unit of volume in a metric system, equal to 1,000 millilitres or 1,000 cubic centimetres (cm^3)
Mole	$6.022 * 10^{23}$ number of molecules
Molecular formula	Shows the number of atoms of each element present in a molecule
Molecular geometry	Shape of a molecule, based on the relative positions of the atoms

Molecular mass	The combined mass (as given on the periodic table) of all the elements in a compound
Molecule	Two or more atoms chemically combined
NGO	Non- governmental organization
Non-recurrent apomixis	The development of embryo takes place from haploid egg cell without fertilization. Such type of apomixis rarely occurs. Generative apospory, haploid parthenogenesis, haploid apogamy and androgamy fall under this category.
Olericulture	Branch of horticulture which deals with cultivation of vegetable crops
One-year-old wood	Term used after leaf fall and during the following growing season to designate shoot or spur growth that developed during the previous growing season.
Oxygen (O_2)	A colourless, odourless, tasteless gas that makes up 21% of our air
Parthenocarpy	Formation of fruits without union of male and female parts
Parthenogenesis	It can be defined as development of embryo from egg cell with or without pollination but without fertilization. Depending upon the ploidy levels of egg cell, parthenogenesis can be haploid (non-recurrent) or diploid (recurrent type) e.g. Mangosteen (*Garcinia mangostana*).
Polyembryony	The phenomenon in which more embryos are present within a single seed is called polyembryony. It may result due to (a) nucellar embryony e.g., Citrus (b) development of more than one nucleus within the embryosac (in addition to the egg embryo during the early stages of development) leading to multiple embryos (e.g.conifers).
Polyembryonic varieties	Varieties which give rise to more than one seedling from one seed
Pasteurizing	Is a process of heat treatment to kill bacteria and help maintain low levels of bacteria within fruit products without major changes in chemistry of the food
Pectin	Is a component found naturally in fruits and can be extracted and used infood processing to form the characteristic gel in jams/jellies
Pectin degrading	Enzyme proteins which facilitate the degradation of pectin. They are used for the the extraction and clarification of fruit juices
pH	Scale for measuring acidity. The pH scale runs from 0 to 14. Values below 7 are acid (sour taste), values above 7 are alkaline (bitter taste)
Plant Propagation	Multiplication of plants by using sexual or asexual plant parts
Polypropylene	A clear glossy film with a high strength and puncture resistance. It has a moderate permeability to moisture, gases and odours

Polythene	A film with a reasonable barrier to moisture, but a relatively high gas permeability
Potassium metabisulphite	Preservative can be added during the processing of dried fruits, jam, fruit leathers, juices, beverages and others to preserve the colour and to extend the shelf-life
Pomology	Branch of horticulture which deals with cultivation of fruit crops
ppm (=parts per million)	Unit of concentration. One ppm is 1 part in 1,000,000. The common
unit mg/l	Is equal to one ppm
Preservatives chemicals	Such as sodium or potassium metabisulphite which are added to the fruit products to prevent the growth of micro organisms and to extend the shelf-life.
Protandry	Stamens mature and anthers shed before the pistil is ready to receive pollen (stigma receptivity does not coincide with pollen viability and maturity); common in insect pollinated plants, more common than protogyny, ensures out crossing by temporal separation *e.g.* walnut, coconut, macadamia, custard apple, sapota and passion fruits.
Protogyny	Stigmas are receptive before anthers release pollen. All monoecious plants and majority of dioecious plants are protogynus *e.g.* banana, plum, pomegranate, avocado, chestnut, pistachionut.
Psuedohermaphroditism:	Plants produce morphologically perfect flowers having both maleand female parts but functionally unisexual (either stamen or pistil is non-functional and behave like either male or female) e.g. grape, plum, pomegranate and persimmon.
Ratio	The relative size of two quantities expressed as the quotient of one divided by the other; the ratio of a to b is written as a:b or a/b
Recurrent apomixis	The embryo sac (female gametophyte) develops from the megaspore mother cell whether meiosis is disturbed (sporogenesis failed) or from adjoining cell (megaspore mother cell disintegrates). The egg cell is diploid and embryo develops directly from the diploid egg cell without fertilization. Generally, somatic apospory, diploid parthernogenesis and diploid apogamy fall under recurrent apomixis. Example: *Rubus* sp. (Raspberry), *Malus hupehensis, Malus sikkimensis, Malus sargentiand Malus*
ICAR	Indian Council of Agricultural Research
RRSS	Regional Research Sub- Station, Raya, SKUAST-Jammu
Refractometer	An instrument which is used to measure the sugar concentration of products

Relative humidity	A measure of the amount of water in the air. It is expressed in a percentage of how much moisture the air could possibly hold
Seed dormancy	Prevention of a seed from germination even when provided with favorable climate and environment
Seed germination	Emersion of embryo from seed
Seed scarification	Breaking, scratching, or softening the seed coat for easy germination
Seed stratification	Seeds of some temperate zone plants which are treated with chilling temperature
Sexual propagation	Union of pollen (male organs) with egg (female organs) to produce of seed
Scaffold limb	Major limbs attached to the main trunk. Shoot: In winter refers to past seasons growth; during the growing season, refers to current seasons growth
Scion	The section of a grafted tree above the graft union
Somatic mutations	These mutations occur in tissues other than the germ track. Most mutations occur somatically, i.e., after the differentiation has set in, when a group of somatic cells is genotypically different from the other cells in the same individual, a somatic mutation may be suspected. The change occurs in the cells of the growing body. Hence the new types of cells are not only heterozygous but form a patch. In meristematic tissues of axillary buds and others a mutation often leads to a batch with new characters. Such changes occur more frequently in polyploidy and heterozygous plants and in individuals which have been grown for long as clones. If propagated vegetatively the mutated parts give rise to new types of plants. This practice is common in horticulture.
Spur	A short shoot that usually terminates in a flower bud
Shelf-life	The amount of time a product can be expected to remain consumable
Skilled labour	Labours having skills about grafting, budding, training, pruning etc
SKUAST-J	Sher-e Kashmir Agricultural Sciences and Technology, Jammu, J&K
SI Unit	Stands for Systeme International d'Unites, a international system which established a uniform set of measurement units
Solute	Substance (solid, liquid, or gas) dissolved in a solution, for example, salt in saltwater
Solution	Mixture of a solid and a liquid where solid never settles out, for example, saltwater
Solvent	Liquid in which something is dissolved, for example, the water in saltwater

Specific heat	The amount of heat it takes for a substance to be raised 1°C
Sodium carbonate	Raising agent aseptic conditions with artificial media under laboratory conditions
Sodium hypochlorite	A chemical compound which is used as a disinfectant. Household bleach usually contains 3 to 6% sodium hypochlorite
Specific gravity	The ratio of the weight of a specific volume of a substance to the weight of the same volume of pure water at 4°C
Tissue Culture Nursery	Technique of asexual propagation by using tissue under aseptic conditions with artificial media under laboratory conditions
Total soluble solids	Content total of all the soluble solids dissolved in water (sugar, salt, protein, acids, etc.)
Terminal	The apex of a shoot or limb
Two-year-wood	Wood one year older than one-year-old wood
Water sprout	An unproductive, fast growing, vertical shoot
Weak acid	Substances capable of donating hydrogen but do not completely ionize in solution
Weak bases	Substances capable of accepting hydrogen but do not completely ionize in solution.

1

Classification of Fruits

1. Fruit : Types and their Classification

Fruit: The botanical definition of fruit is a seed-bearing part of a flowering plant or tree that can be eaten as food

or

Fruits are matured ovaries plus any associated flower parts and contain the seeds of the plant. The ovary may be subdivided into two or more carpels (then termed compound ovary) each bearing one to many ovules. Individual carpels develop into sections of a whole fruit, as with citrus where each familiar segment of the fruit represents one matured carpel. If the ovary is not subdivided, then it is termed simple. The ovules will mature into seeds if fertilized. Fruit has three parts.

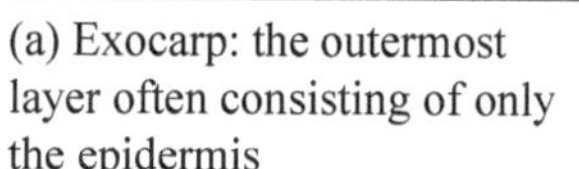

(a) Exocarp: the outermost layer often consisting of only the epidermis	(b) Mesocarp or middle layer: which varies in thickness	(c) Endocarp: which shows considerable variation from one species to another

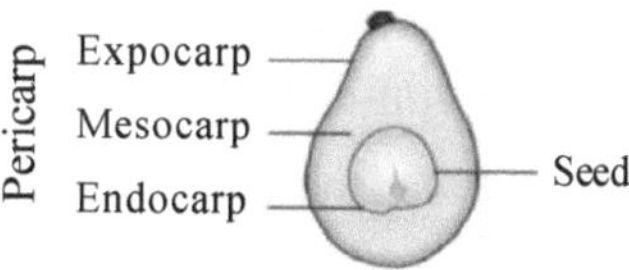

Figure 1

The exocarp (middle), generally, becomes the fruit peel or skin, the mesocarp becomes the fruit flesh and the endocarp (innermost) becomes the innermost part of the flesh or a specialized tissue surrounding the seed(s), like a pit. These layers may develop into distinct parts of the fruit. In many cases, however, the three layers are indistinguishable and the term "pericarp" is applied to denote all ovarian tissues surrounding the seed(s). Fruits are matured ovaries and contain the seeds of the plant. Ovaries sometimes have distinct layers of tissues, as labelled above, which develop into distinct structures of the fruit as shown for a drupe. Not all fruits have such clearly demarcated tissues. All fruits may be classified into three major groups on the basis of the number of flowers and the number of ovaries involved in their formation. The following outline includes most of the common types of fruits. A simple way of fruit classification is provided here-

Types of fruits

1. Based on the origin

Horticultural fruits may be divided into two types-true fruits & false fruits. True fruits are those which are developed wholly from tissues of a single ovary. Such fruits are called simple fruits, like peach, mango, plum and orange. An aggregate fruit is formed when several ovaries of a single flower develop independently but remain attached to their common stem, as in the case of the raspberry. A false fruit, viz., apple, pear and strawberry, it is one that is composed of tissues in addition to those of the ovary. A multiple fruit is a false fruit that is formed when several adjacent flowers and their tissues unite to form a single fruit such as pineapple.

A. True
   (Ovary wall only)

B. Accessory
   (Ovary wall plus other parts)

i. Receptacle

ii. Bracts

iii. Perianth

2. Based on fruit morphology

A. **Simple fruit**- Simple fruits develop from a single matured ovary in a single flower. Accessory fruits have some other flower part united with the ovary.

B. **Fleshy fruits**- Pericarp fleshy at the time of maturity

C. **Berry-** Consisting of one or more carpels with one or more seeds, the ovary wall is fleshy banana, Papaya, Grape, Sapota and Avocado

D. **Modified berry**

i. Balusta: Pomegranate

ii. Amphisarca: Wood apple and Bael

iii. Pepo: (an accessory fruit), a berry with a hard rind, the receptacle partially or completely enclosing the ovary i.e. Water melon

iv. Pome: (an accessory fruit), derived from several carpels, receptacle and outer portion of pericarp fleshy, inner portion of pericarp is papery or cartilaginous, forming a core i.e. Apple, Pear and Loquat

v. Drupe (Stone): A stonefruit, derived from a single carpel and containing (usually) one seed. Exocarpa thin skin i.e. Mango, Pear and Plum

vi. Hesperidium: A specialized berry with a leathery rind i.e. Citrus

vii. Nut: Cashew, Litchi, Walnut and Rambutan

viii. Capsule: Aonla and Carambola

Dry fruits, pericarp dry at the time of maturity

A. **Dehiscent fruits,** those which dehisce or split or open when fully mature

a. **Follicle,** composed of one carpel and splitting along a single suture

b. **Legume,** composed of a single carpel and splitting along two sutures

c. **Capsule,** composed of several carpels and opening at maturity in one of four ways:

i. Along the line of carpel union (septicidal dehiscence)

ii. Along the middle of each carpel (loculicidal dehiscence)

iii. By pores at the top of each carpel (poricidal dehiscence)

iv. Along a circular, horizontal line (circumscissle dehiscence)

B. **Indehiscent fruits,** those which do not split or open at maturity

a. **Achene or akene,** a one-seeded fruit with the seed attached to the fruit at one point only

b. **Caryopsis or grain,** a one-seeded fruit in which the seed is firmly attached to the fruit at all possible points

c. **Samara,** a one- or two-seeded fruit with the pericarp bearing a wing like outgrowth. A modified achene

d. **Schizocarp,** consisting of two carpels which at maturity separate along the midline into two one-seeded halves, each of which is indehiscent.

e. **Loment,** having several seeds, breaking into one-seeded segments at maturity.

f. **Nut**, a hard, one-seeded fruit, generally formed from a compound ovary, with the pericarp hard throughout.

C. **Aggregate fruits**: Aggregate fruits consist of a number of matured ovaries formed in a single flower and arranged over the surface of a single receptacle. Individual ovaries are called fruitlets, Eteario of berries, Custard apple and Raspberry

D. Multiple fruit: Multiple fruits consist of the matured ovaries of several to many flowers more or less united into a mass. Multiple fruits are almost invariably accessory fruits

Syconus-Fig. Sorosis - Jackfruit, Pineapple and Mulberry

E. Siliqua: composed of two carpels which separate at maturity, leaving a persistent partition between them

2. Angiosperms

Plants can be identified by observing certain distinguishing morphological and bio-chemicals characteristics. Some plants are closely related, which is shown by the similarity of their leaves, canopy structures, flower structures, fruit shape and size. These plants are placed into a specific plant family and in each family there are members that are more closely related than others. This relationship is demonstrated by the similarity of basic morphological traits like leaf shape, leaf colour or fruit shape and size arrangement. These plants are placed in a group called a ***genus***. Members of a plant genus are again subdivided, according to their similar morphological characteristics, into a grouping called a ***species***. Botanical classification based on botanical relationship with genomes

i. Biggest family: Rutaceae

ii. Smallest family: Passifloraceae

3. Based on Rate of Respiration

a. Climacteric Fruits: Mango, Banana, Sapota, Guava, Papaya, Apple, Fig, Peach, Pear, Plum and Annona

b. Non climacteric Fruits: Citrus, Strawberry, Grape, Pomegranate, Pineapple, Litchi, Ber, Jamun, Cashew and Cherry

(Climacteric fruits produce much larger amount of ethylene than the non climacteric fruits)

4. Based on Light Requirement

a. *Heliophytes:* Plants which grow in open sunny situation

b. *Sciophytes:* Plants which grow in shade

i) Facultative sciophytes: Plants which grow in shade but not so optimally

ii) Obligate sciophytes: Plants which always grow in shade

c. *Facultative heliophytes:* Plants which grow in shade

d. *Obligate heliophytes:* Plant which always grow under many situation

Table 1: Botanical name, family, fruit types, edible portion, chromosome no & origin of fruits

S. No.	English Name	Common Name	Botanical Name	Family	Fruit types	Edible portion	Ch No. 2n=	Origin
1.	Aonla	Amla	*Emblica officinalis* Gaertn	Euphorbiaceae	Capsule (Modified Berry)	Mesocarp and Endocarp	28	S-E Asia
2.	Almond	Badam	*Prunus dulcis* Mill	Rosaceae	Drupe	Cotyledon	16	Central Asia
3.	Apple	Seb	*Malus pumila* Mill	Rosaceae	Pome	Fleshy thalamus	34	S-W Asia
4.	Apple (Crab apple)	Chhota seb	*Malus baccata* Borgen,	Rosaceae	Pome	Fleshy thalamus	34	S-W Asia
5.	Apricot	Khurmani	*Prunus armeniaca* L.	Rosaceae	Drupe	Epicarp and Mesocarp	16	China
6.	Atemoya	Lakshman phal	*Annona atemoya* Hort.	Annonaceae	Etaerio of berry	Pericarp	14	Man Made hybrid
7.	Avocado	Avocado	*Persea americana* Mill.	Lauraceae	Berry	Pericarp	24	C. America
8.	Bael	Bil	*Aegle marmelos* Correa.	Rutaceae	Amphisarica (Berry)	Succulent	18	India
9.	Banana	Kela (edible fruit)	*Musa sapientum* L.	Musaceae	Multi seed Berry	Mesocarp and endocarp	22,33, 44	Indo malaya
10.	Banana	Kela (Plantain cooking type)	*Musa paradisiaca* L	Musaceae	Berry	Mesocarp and endocarp		
11.	Breadfruit	Valayeti phanas	*Artocarpus artilis* Fos.	Moraceae	Sorosis	Bracts/ Perianth/ seed	56	W. Ghats of India
12.	Bullock's heart	Ram phal	*Annona reticulata* L	Annonaceae	Etaerio of berry	Pericarp	14	W.India

13.	Calamondin	China Sangtra	*Citrus madurensis* Lour.	Rutaceae	Hesperidium	Juicy placental	-	-
14.	Cashew nut	Kaju	*Anacardium occidentale* L.	Anacardiaceae	Nut	Cotyledon and fleshy peduncle	42	Brazil
15.	Cattley guava	Strawberry guava	*Psidium cattleianum* L.	Myrtaceae	Berry	Thalamus and pericarp	-	-
16.	Cherimoya	Hanuman phal	*Annona cherimoya* Mill.	Annonaceae	Etaerio of berry	Pericarp	14	Bolivia
17.	Cherry (Sour)	Sour cherry	*Prunus cerasus* L.	Rosaceae	Drupe	Mesocarp Endocarp	16	Asia minor
18.	Cherry (Sweet)	Sweet cherry	*Prunus avium* L.	Rosaceae	Drupe	Mesocarp Endocarp	16	Asia minor
19.	Chinese jujube	Ber	*Ziziphus jujube* Lamk.	Rhamnaceae	Drupe	Pericarp	48(4x)	S-W Asia
20.	Citron	Turange	*Citrus medica* L.	Rutaceae	Hesperidium	Juicy placental	18	-
21.	Cleopatra	Billi kichili	*Citrus reshni* Tanaka	Rutaceae	Hesperidium	Juicy placental	18	-
22.	Coconut	Khopa or Nariyal	*Cocos nucifera* L.	Palmaceae	Drupe	Endosperm	-	-
23.	Custard apple	Sita phal	*Annona squamosa* L.	Annonaceae	Etaerio of berry	Pericarp	14	W-India
24.	Date	Khajoor	*Phoenix dactylifera* L.	Aracaceae	Drupe	Pericarp	36	W-Asia
25.	Date (Wild)	Khajoor	*Phoenix sylvestris* Roxb.	Palmaceae	Drupe	Pericarp	36	W-Asia
26.	Feijoa	Pineapple guava	*Feijoa sellowiana* Berg.	Myrtaceae	Berry	Thalamus and pericarp	22	T.America
27.	Fig	Anjeer	*Ficus carica* L.	Moraceae	Syconus	Fleshy receptacle	26	W-Asia

28.	Kiwi	Kiwi fruit	*Actinidia deliciosa*	Actinidiaceae	Berry	Mesocarp and Placenta	58	C.America
29.	Grape (American)	Angoor	*Vitis labrusca* Bailey	Vitaceae	Berry	Pericarp and placentae	38	W-Asia
30.	Grape (European)	Angoor	*Vitis vinifera* L.	Vitaceae	Berry	Pericarp and placentae	38	W-Asia
31.	Guava	Amrood	*Psidium guajava* L.	Myrtaceae	Berry	Thalamus and pericarp	22	T.America
32.	Indian jujube	Ber	*Zizyphus mauritiana* Lamk.	Rhamnaceae	**Drupe**	Pericarp	48(4x)	S-W Asia
33.	Jackfruit	Katehal	*Artocarpus heterophyllus* Lam.	Moraceae	Sorosis	Bracts/perian th/ seed		
34.	Jambolan	Jamun	*Syzygium cuminii* Skeels	Myrtaceae	Berry	Mesocarp Epicarp	40	India
35.	Japanese summer grape fruit	Japani grapefruit	*Citrus natsudaidai* Hayat.	Rutaceae	Hesperidium	Juicy placental	-	-
36.	Kainth	Kainth	*Pyrus pashia* Buch-Ham.	Rosaceae	Pome with grit cells	Fleshy thalamus	34	W.China
37.	Karonda	Karonda	*Carissa carandas L.*	Apocynaceae	Berry	Epicarp and Mesocarp	12	India
38.	Kharna khatta	Kharna khatta	*Citrus karna* Raf.	Rutaceae	Hesperidium	Juicy placental	18	-
39.	Kumquat (oval)	Japani narangi	*Fortunella margarita* Swingle	Rutaceae	Hesperidium	Juicy placental	18	-
40.	Kumquat (round)	Japani narangi	*Fortunella japonica* Swingle	Rutaceae	Hesperidium	Juicy placental	18	-

41.	Lasoda	Lasoora	*Cordia myxa* Roxb.	Boraginaceae	-	Mesocarp Native of northwestern India	-	-
42.	Lemon	Baramasi lemon	*Citrus limon* Burm.	Rutaceae	Hesperidium	Juicy placental	18	-
43.	Lemon	Galgal	*Citrus limon* Burm.	Rutaceae	Hesperidium	Juicy placental	-	-
44.	Litchi	Litchi	*Litchi chinensis* Sonn.	Sapindaceae	Nut	Fleshy aril	30	S. China
45.	Loquat	Loquat	*Eriobotrya japonica* Lindl.	Rosaceae	Pome	Fleshy thalamus	32	E. China
46.	Mandarin	Sangtra	*Citrus reticulata* Blanco.	Rutaceae	Hesperidium	Juicy placental	18	S-E Asia
47.	Mandarin (Satsuma)	Sangtra	*Citrus unshiu* Marc.	Rutaceae	Hesperidium	Juicy placental	18	-
48.	Mango	Aam	*Mangifera indica* L	Anacardiaceae	Drupe	Mesocarp	40	S-E Asia
49	Mangosteen	Mangosteen	*Garcinia mangostana* L.	Guttiferae	Berry	Endocarp	24	S-E Asia
50.	Mulberry (Black) Mulberry (White)	Shehtoot Shehtoot	*Morus nigra* L. *Morus alba* L.	Moraceae Moraceae	Syncarpous/ Aggregate	Mesocarp	308	China
51.	Olive	Zaitoon	*Olea europaea* L.	Oleaceae	Drupe	Epicarp and Mesocarp		
52.	Papaya	Papita	*Carica papaya* L	Caricaceae	Berry	Mesocarp	18	T.America
53.	Passion fruit	Passion fruit	*Passiflora edulis* Sims.	Passifloraceae	Berry	Mesocarp	18	Brazil
54.	Peach	Aru	*Prunus persica* L.	Rosaceae	Drupe	Mesocarp	16	China

55.	Pear (European)	Nashpati	*Pyrus communis* L.	Rosaceae	Pome	Fleshy thalamus	34	W.Asia
56.	Pear (Sand pear)	Sand pear	*Pyrus pyrifolia* L.	Rosaceae	Pome	Fleshy thalamus	34	W.Asia
57.	Pecan nut	Pecan nut	*Carya illinoinensis* Koch.	Juglandaceae	Nut	Seed cotyledons	32	C.Asia
58.	Pectinifera	Pectinifera	*Citrus pectinifera* Tanaka	Rutaceae	Hesperidium	Juicy placental	-	-
59.	Persimmon	Japani phal	*Diospyros kaki* L.	Ebenaceae	Berry	Mesocarp	90 (6x) 60 (4x)	China
60.	Phalsa	Phalsa	*Grewia asiatica* D.C.	Tiliaceae	Drupe	Epicarp and Mesocarp	36	India
61.	Pilu	Pilu	*Salvadora oleioides* Decae.	Salvadoraceae	Drupe	-	-	-
62.	Pineapple	Ananas	*Ananas comosus* Merr.	Bromeliaceae	Sorosis	Aggregate berry	50,75, 100	-
63.	Pistachio nut	Pista	*Pistacia vera* L.	Anacardiaceae	Nut	Cotyledons	30	Iran
64.	Plum (English)	Alubokhara	*Prunus bokhariensis* Schneid.	Rosaceae	Stone	Mesocarp Epicarp		China
65.	Plum (European)	Alucha	*Prunus domestica* L.	Rosaceae	Stone	Mesocarp Epicarp	48	China
66.	Plum (Japanese)	Alucha	*Prunus salicina* Lind\.	Rosaceae	Stone	Mesocarp Epicarp	16	China
67.	Pomegranate	Anar	*Punica granatum* L.	Punicaceae	Modified berry (Balausta)	Juicy testa	-	-
68.	Pomelo	Grapefruit	*Citrus paradisi* Macf.	Rutaceae	Hesperidium	Juicy placental	18	-
69.	Pummelo	Chakotra	*Citrus maxima* Osbeck	Rutaceae	Hesperidium	Juicy placental	18	-

70.	Quince	Beedana	*Cydonia oblonga* Mill.	Rosaceae	Pome	Fleshy thalamus	34	IranTurkey
71.	Rambutan	Rambutan	*Naphelium jappaceum*	Sapindaceae	Berry	Aril (Fleshy testa)	22	Malayan
72.	Rangpur lime	Sylhet lime	*Citrus limonia* Osbeck	Rutaceae	Hesperidium	Juicy Placental	18	-
73.	Raspberry	Rasbhari	*Rubus idaeus* L.	Rosaceae	Etaerio of berry	Thalamus	14	N.America
74.	Rose apple	Gulab jamun	*Eugenia jambos* L.	Myrtaceae	Berry	Pericarp with thalamus fused	28	S.Brazil
75.	Rough lemon	Jatti khatti	*Citrus jambhiri Lush*	Rutaceae	Hesperidium	Juicy Placental	18	-
76.	Sapota	Chiku	*Achras sapota* L.	Sapotaceae	Berry	Mesocarp and endocarp	18	-
77.	Sour lime	Nimbu	*Citrus aurantifolia Swingle*	Rutaceae	Hesperidium	Juicy placental	18	-
78.	Sour orange	Khatta	*Citrus aurantium* L.	Rutaceae	Hesperidium	Juicy Placental	18	-
79.	Strawberry	Strawberry	*Fragaria chiloensis Duch*	Rosaceae	Etaerio of achenes	Fleshy thalamus	56 (8x)	-
80.	Suranam cherry	Pitanga	*Eugenia uniflora* L.	Myrtaceae	Berry			
81.	Sweet lime	Mitha	*Citrus limettioides Swingle*	Rutaceae	Hesperidium	Juicy Placental	18	-
82.	Sweet orange	Malta	*Citrus sinensis L.*	Rutaceae	Hesperidium	Juicy Placental	18	-
83.	Tahiti lime	Tahiti ni mbu	*Citrus latifolia Tanaka*	Rutaceae	Hesperidium	Juicy Placental	18	-
84.	Tamarind	Imli	*Tamarindus indica* L.	Leguminosae	Pod	Pulp or Mesocarp		
85.	Trifoliate orange (deciduous orange)	Tinpattia	*Poncirus trifoliata* L.	Rutaceae	Hesperidium	Juicy Placental	18	-
86.	Walnut	Akhrot	*Juglans regia* L.	Juglandaceae.	Drupe	Cotyledon	32	-
87.	Wood apple	Hathi seb	*Feronia limonia* L.	Rutaceae	Amphisarca (berry)	Succulent	18	India

5. Based on Water Requirement

a. Hydrophytes: Those plants which grow partially in fully submerged water conditions such as banana

b. Mesophytes: Plants which grow in situation where water is neither in abundant nor scarce i.e., Papaya, mango, guava, apple, peach and pear etc.

c. Xerophytes: Plants which grow in extremely scarce conditions *viz.*, Pilu and kair

6. Based on Photo Periodic Requirement

a. Short day plant
(Light period of 12 hr or less) such as strawberry

b. Long day plant
(Light period of 12 hr or more) such as banana

7. Based on Branching Habit

a. Heliotropic branching habit: Plants with branches emerging high-up on the stem and growing upward. For example Hog Plum

b. Geotropic branching habit: Plants with branches emerging in near proximity to ground and growing parallel to the ground surface. For example Banyan.

8. Based on Longevity

a. Very Long longevity: >100 yrs - Datepalm, Coconut, Ber, Walnut and Arecanut

b. Long longevity: 50-100 yrs - Mango, Bael and Tamarind

c. Medium: 10-50 yrs - Litchi, Guava, Phalsa, Karonda, Citrus and Pomegranate

d. Short: Pineapple, Strawberry, Banana and Papaya

9. Based on Consumers Preference or Weight of Fruits

a. Very light: 50-100 gm Example Grape, Karonda, Cherry, Strawberry, Ber and Banana

b. Light: 100-150 gm Example Sapota, Kinnow and Pomegranate

c. Light medium: 150-300 gm Example Mango and Avocado

d. Heavy: 500 g - 1 kg Example Bread fruit, Papaya, bael and Pineapple

e. Very heavy: >5kg Jackfruit

10. Inforescence

Fruit set is the process in which flowers become fruit and potential fruit size is determined. It occurs after pollen is released from the flower parts (anthers), lands on receptive female flower parts (stigmas), produces a pollen that grows to the ovule, and fertilizes the egg within the ovule. The fertilized egg forms the seed, which induces the surrounding (pericarp) tissue to grow and form a berry. The floral morphology of a fruit crop is important in determining the mode of pollination (i.e., wind, gravitation force, water or insect) and type of fruit that will arise when the ovary gets mature. A complete flower is one possessing all four fundamental appendages: sepals, stamens, petals and pistils. An incomplete flower lacks one or more of these features. The position of the base of the pistil, or ovary, with respect to the other three appendages is important in identification and also partially determines what the fruit type will be. The two most common positions are "inferior" (epigynous) and "superior" (hypogynous). A third possibility is "half-inferior" (perigynous), as found in the stone fruits (*Prunus* spp.). A perigynous ovary sits within a hypanthium or floral cup. Different fruit types arise from the flowers shown, partly as a result of the difference in ovary position. Flowers are borne on structures called inflorescence, which is a collection of individual flowers arranged in a specific order or form.

a. Spike b. Catkin c. Raceme d. Corymb e. Umbel f. Compound umbel g. Cyme h. Panicle i. Head and j. Solitary flower

11. Based on Type of Pollination

a. Autogamy or self-fertilization refers to the fusion of two gametes that come from one individual. It is mainly observed in the form of self-pollination, a reproductive mechanism employed by many flowering plants

b. *Cleistogamy:* It is a type of automatic self-pollination of certain plants that can propagate by using non-opening, self-pollinating flowers Example: Grape and Sapota

c. *Homogamy:* It can be used as a form of choosing a mate based on characteristics that are wanted in a sexual partner such as phalsa, apricot, citrus etc.

d. *Parthenocarpic:* Fruit development without fertilization. The fruit resembles a normally produced fruit but it is seedless. Pineapple, banana, cucumber, grape, orange, grapefruit, persimmon and breadfruit cultivars are exemplify naturally occurring. In Parthenocarpy, the ovary is stimulated even without pollination and thus fruit development begins without fertilization. This is common in plants that have no ovary or plants that have lost their ability to reproduce sexually due to mutation.

Apomixis

It is the formation of seeds without fertilization. In a natural flow of biological processes, pollination is the first step in the formation of a fruit and seed. But, in this case, there is no meiotic division and fertilization of the gametes to form a zygote. It is two types:

1. *Sporophytic* – In this kind, apomixis occurs from the diploid sporophyte

2. *Gametophytic* – In this kind, apomixis occurs from the haploid gametophyte

Table 2 : Fruit crops and their pollination requirements

S.No.	Fruit crop	Status
1.	Aonla	Self-pollinated /Cross-pollinated
2.	Bael	Self-pollinated /Cross-pollinated
3.	Banana	Self-pollinated
4.	Ber	Cross-pollinated
5.	Carambola	Self-pollinated /Cross-pollinated
6.	Cashew	Cross-pollinated
7.	Citrus	Self-pollinated /Cross-pollinated
8.	Coconut	Cross-pollinated
9.	Custard apple	Cross-pollinated
10.	Date palm	Self-pollinated /Cross-pollinated
11.	Fig	Parthenocarpic
12.	Grapes	Self-pollinated
13.	Guava	Self-pollinated
14.	Jackfruit	Self-pollinated /Cross-pollinated
15.	Jamun	Cross-pollinated
16.	Karonda	Self-pollinated
17.	Lasara	Self-pollinated
18.	Litchi	Cross-pollinated
19.	Loquat	Self-pollinated /Cross-pollinated
20.	Mango	Cross-pollinated
21.	Mangosteen	Parthenocarpic
22.	Mulberry	Cross-pollinated
23.	Papaya	Self-pollinated /Cross-pollinated
24.	Phalsa	Self-pollinated
25.	Pine apple	Cross-pollinated
26.	Pomegranate	Self-pollinated /Cross-pollinated
27.	Sapota	Cross-pollinated
28.	Almond	Self-pollinated /Cross-pollinated
29.	Apple	Self-pollinated /Cross-pollinated
30.	Apricot	Self-pollinated
31.	Cherry	Cross-pollinated
32.	Peach	Cross-pollinated
33.	Pear	Cross-pollinated
34.	Plum	Cross-pollinated
35.	Strawberry	Self-pollinated /Cross-pollinated
36.	Walnut	Self-pollinated

12. Leaf Morphology

Angiosperm leaf types

The location of the bud determines whether the leaf is simple, compound and Variation in Compound Leaves. In the case of the single leaf the bud is found in the axil of the leaf and stem

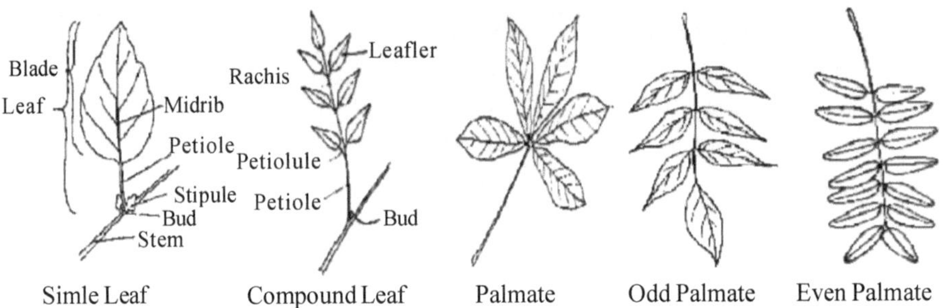

Figure 2 : Simple Leaf, Compound Leaf and Variation in Compound Leaves

There are five categories, that classify plants into groups and it assists in eliminating many plants from consideration in the process of positive identification.

a. Leaf Shapes

Involvement of the following images with the terms will help us alleviate the burden of strict terminology. This also applies to leaf bases, margins and apices.

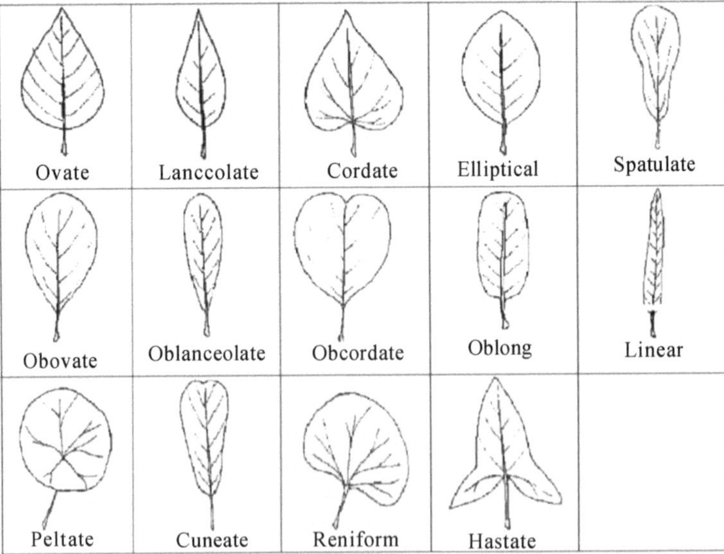

Figure 3

b. Leaf basis

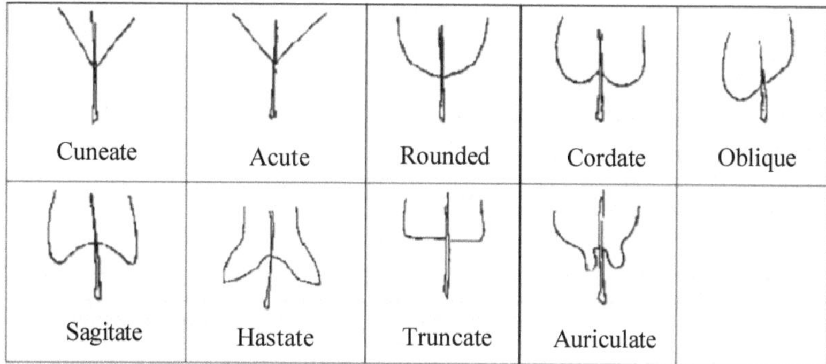

Figure 4

c. Leaf margins

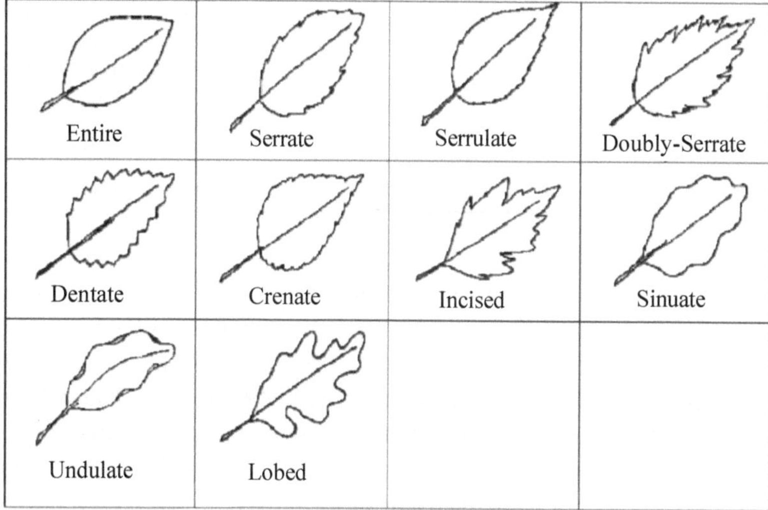

Figure 5

d. Leaf apices

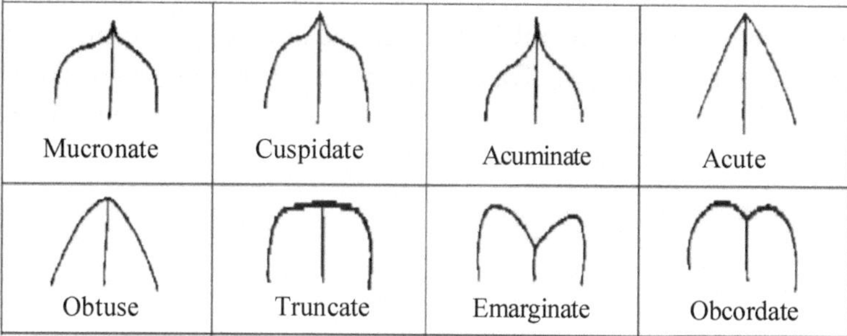

Figure 6

13. Horticultural Zones of India and their Classification as Plant Behavior

i. Broadly, India can be divided into three regions, Tropical, Subtropical and Temperate (hilly region).

ii. Every region is category differences due to rainfall, topography, humidity, altitude and sunshine hour etc.

iii. Choice of the crops can be made and development is planned.

Generally India can be divided into three regions, Tropical, Subtropical and Temperate

Temperate	N.W and N.E Subtropical	Tropical zones
Kashmir, Himachal Pradesh, North Uttaranchal, Sikkim and part of Arunachal Pradesh	Punjab, Haryana, Rajasthan, Central Uttar Pradesh and North M.P., Bihar, Jharkhand, Assam,Meghalaya, Nagaland and Manipur.	South Madhya Pradesh, Chattisgarh, Gujarat, Maharashtra, Orissa and West Bengal. Southern tropical: Karnataka, Andhra Pradesh and Tamil Nadu. Coastal tropical humid: Konkan, Goa, Kerala, Western Ghats, Eastern Ghats in Tamil Nadu, Andhra Pradesh and Orissa.

The regions where subtropical fruits are produced are between the true tropics, where frost never occurs, and the temperate regions, where normally the temperature falls below freezing and stays below for a considerable part of the winter season. In these subtropical, or intermediate, zones the temperature occasion-ally goes below freezing but not as a rule below 25°F. The influence of large bodies of water and the protection of mountain ranges, or planting where topography gives good air drainage, sometimes extend this type of region as "islands" into territory beyond the usual subtropical regions. The types of fruits grown in subtropical regions merge into those grown in the true tropics; and no hard-and-fast division can be drawn on the basis of fruit types except that forms possessing resistance to light freezing temperature are of major importance in the subtropics. Diverse types are cultivated, but the orange, grapefruit and lemon are the most commonly known. Dates, figs and olives are also important. Many true tropical fruits such as the avocado, mango, papaya and even the pineapple are also grown on a commercial scale in the sub-tropical regions.

Broad outlines of the horticultural crops grown in each zone are as follows

S.No.	Tropical Zone (100 – 300 msl)	Sub-tropical Zone (300-1100 msl)	Temperate Zone (1100-2000 msl)
1	Litchi	Litchi	Plum
2	Papaya	Lemon	Peach
3	Lemon	Orange	Pear
4	Orange	Guava	Apricot
5	Pineapple	Grape fruit	Chesnut
6	Banana	Pineapple	Apple
7	Guava	Ginger	Walnut
8	Arecanut	Apple	Persimmon
9	Black Pepper	Walnut	Cherry
10	Betle Lea	Mango	Strawberry
11	Sapota	Aonla	Pecanut
12	Mango	Sapota	Almond
13	Aonla	Ber	Quince
14	Ber	Pomegranate	Pomegranate
15	Strawberry	Strawberry	

a. These regions are characterized by different agro climatic features

i. Annual rainfall and humidity

ii. High solar radiation incidence (450-500 cal per cm^2/day) and high wind velocity (20km/hour) result in a high potential evapotra spiration (6mm/day) and high mean humidity index (74-78%).

iii. The soils being windblown, have 85% sand and low organic matter (0.1-0.45%) with poor water holding capacity (25-28%) and high infiltration rate (9cm/hr).

iv. The ground water resources are inadequate.

b. Fruit growing regions in India

1. Temperate zone: Himachal Pradesh, Jammu & Kashmir, Uttarakhand, Arunachal Pradesh, parts of Nagaland, Nilgiris and Pulney hills in Tamil Nadu.
2. North Western subtropical zone: Punjab, Haryana, Rajasthan, Madhya Pradesh.
3. North Eastern subtropical zone: Bihar, Assam, Meghalaya, Tripura, part of Arunachal Pradesh and West Bengal.
4. Central tropical zone: Madhya Pradesh, Maharashtra, Gujarat, Orissa, West Bengal, Andhra Pradesh and Karnataka.
5. Southern tropical zone: Karnataka, Andhra Pradesh, Tamil Nadu and Kerala.

6. Coastal tropical humid zone: Coast of Maharashtra, Kerala, Andhra Pradesh, Tamil Nadu, Orissa, West Bengal, Tripura and Mizoram, Gujarat along sea and the Indian Islands.

The ICAR, New Delhi, has recognized eight agro climatic zones for effective land useplanning.

S.No.	Agro climatic	Region Status
1.	Humid western	Himalayan Region J&K, HP, Kumaon and Garhwal in Uttarakhand
2.	Humid Bengal –	Assam Region West Bengal & Assam
3.	Humid Eastern	Himalayan Region Bay Islands, Arunachal Pradesh, Nagaland, Manipur, Mizoram, Tripura, Sikkim, Meghalaya & Andaman and Nicobar Islands.
4.	Sub-humid Sutlez	Ganga Alluvial plains Punjab, Delhi, UP plains and Bihar.
5.	Sub-humid to Humid Eastern and south Eastern	IslandsEastern Madhya Pradesh, Orissa and Bihar.
6.	Arid western plains Haryana, Rajasthan, Gujarat, Dadra and Nagar	Haveli and Daman and Diu
7.	Semi-arid Lava plateausand central Islands.	Maharastra, Western Central Madhya Pradesh and Goa.
8.	Humid to Semiarid	Western Ghats Karnataka, Tamil Nadu, Kerala, Pondichery and Lakshadweep Islands.

During 1985-90, the Planning commission accepted 15 broad agro climatic zones based on physiographic and climate for effective planning.

These zones are:

1. Western Himalayan Region.
2. Eastern Himalayan Region.
3. Lower Gangetic plains Region.
4. Middle Gangetic plains Region.
5. Upper Gangetic plains Region.
6. TransGangetic plains Region.
7. Eastern Plateau and Hills Region.
8. Central Plateau and Hills Region.
9. Western Plateau and Hills Region.
10. Southern Plateau and Hills Region.
11. East coast Plains and Hills Region.
12. West coast Plains and Ghats.
13. Gujarat plains and Hills Region.
14. Western Dry Region.
15. Island Region.

Different agro- climatic region showing in map

Importance of fruits for human nutrition

Fruits have important role in balanced diet, they provide not only energy rich food but also provide vital protective nutrients/elements and vitamins. Since, most of Indians are vegetarians, the incorporation to eat fresh fruits and their produce in daily diet is essential for good health. With the growing awareness and inclination towards vegetarianism worldwide, the fruit crops are gaining tremendous importance.

Functions of fruits in human body

1. Fruits are provide palatability, taste, improves appetite and provide fiber.
2. They neutralize the acids produced during digestion of proteins and fatty acids.

Table 13 : Approximate nutrient composition (per 100 g) of convention and non-convention fruits

a. Convention fruits

Fruit crop	Calories (kcal)	Carotene Vit.A(IU)	Vit.B1 (mg)	Vit.B2 (mg)	Niacin (mg)	Vit.C (mg)	Ca (mg)	P (mg)	Fe (mg)	K (mg)
Almond	655	trace	240	-	2.5	-	0.23	0.49	3.5	-
Apple	56	Trace	120	30	0.2	2	0.01	0.02	1.7	144
Apricot	201	98	217	-	2.2	trace	-	-	0.04	-
Banana	153	trace	150	30	0.3	1	0.01	0.05	0.4	396
Grape	45	15	40	10	0.3	3	0.03	0.02	0.4	0.15
Grape fruit	45	-	120	20	0.3	31	0.03	0.03	0.2	-
Guava	66	trace	30	30	0.2	299	0.01	0.04	1.0	-
Lemon	57	trace	89	4	0.1	60.0	0.07	0.01	2.3	-
Lime	59	26	20	-	0.1	63.0	0.09	0.02	0.3	-
Litchi	42	14	87	122	-	trace	0.01	0.31	0.03	-
Mandarin	49	50	120	60	0.3	68	0.05	0.02	0.1	-
Mango	50	4800	40	50	0.3	13	0.01	0.02	0.3	-
Papaya	40	2020	40	250	0.2	48	0.01	0.01	0.4	-
Peach	38	trace	20	1	0.2	1	0.01	0.03	1.7	-
Pear	47	14	20	30	0.2	trace	0.01	0.01	0.7	-
Pineapple	50	60	-	120	-	63	0.02	0.01	0.9	-
Plum	40	230	120	30	0.3	1	0.02	0.02	0.5	-
Pomegranate	65	-	-	100	-	16	0.01	0.07	0.3	-
Pummelo	45	200	30	-	0.2	20	0.03	0.03	0.1	-
Strawberry	44	-	30	-	0.2	52	0.03	0.03	1.8	-
Sweet orange	53.8	-	-	-	-	50.0	42.0	23.0	0.40	177
Walnut	687	10	450	-	1.6	-	0.10	0.38	4.8	
b. Non convention fruit crops										
Avocado	215	-	-	-	-	13	0.01	0.08	0.7	-
Bael	129	186	12	1191	0.9	15	0.09	0.05	0.3	-
Ber	55	70	-	-	-	160.0	0.03	0.03	0.8	-
Blueberry	62.4	-	-	-	-	22.0	10.0	9.1	0.74	65
Breadfruit	79	15	-	-	-	-	0.04	0.03	0.5	-
Cape Gooseberry	55	-	-	-	-	49	0.01	0.06	1.8	-
Cherry	60.2	-	-	-	-	12.0	8.0	7.0	-	114
Currant	45.0	-	-	-	-	36.0	29.0	27.0	0.91	238
Custard apple	105	trace	-	-	-		0.02	0.04	1.0	-
Fig	75	270	-	50	0.6	2	0.06	0.03	1.2	194
Jackfruit	84	540	30	-	0.4	-	0.02	0.03	0.5	-
Jamun	83	-	-	-	-	10	0.02	0.01	1.0	-
Karonda	364	-	-	-	-	-	0.16	0.06	39.1	-
Mangosteen	60	-	-	-	-	-	0.01	0.02	0.02	-
Passion fruit	93	90	-	-	-	-	0.01	0.06	2.0	-
Wood apple	97	-	-	170	-	-	0.13	0.11	0.6	
Persimmon	81	1710	-	-	-	-	0.01	0.01	0.3	-
Raspberry	40.2	-	-	-	-	25.5	40.0	44.0	1.00	170
Timru	112	361	0.04	-	-	1.0	60.0	-	0.5	-
Aonla	59	-	30	-	0.2	600	0.05	0.02	1.2	-

3. They improve the general immunity of human body against diseases, deficiencies etc.
4. They are the important source of vitamins and minerals for used in several bio-chemical reactions occur in body.

14. National Fruits of Different Countries

S.No.	Name of fruit	Country
1.	Winter guava and Summer mango	Pakistan
2.	Damson plum	Serbia
3.	Jack fruit	Bangladesh
4.	Avocado	Mexico
5.	Kiwi fruit	New Zealand
6.	Apricot	Armenia
7.	Jack fruit	Bangladesh
8.	Mango	India
9.	Filipino mango	Phillipines
10.	Japanese persimmon/ litchi	Japan
11.	Apple	Austria, England
12.	Lady's finger	Cambodia
13.	Fuzzy kiwi	China
14.	Red durian	Indonesia
15.	Rambutan/ Papaya	Malaysia
16.	Durian	Singapore
17.	Mangosteen	Thailand
18.	Banana	Central African republic
19.	Borojo, Curuba	Columbia
20.	Ackee	Jamaica
21.	Banana	Central African republic
22.	Borojo, Curuba	Columbia

15. Based on Rate of Ethylene Production

Level of ethylene	Rate (μl/kg/hr)	Example
Very low	<0.1	Grapes and Citrus
Low	0.1-1.0	Pineapple
Medium	1-10	Mango, Banana, Guava, Fig , Litchi
High	10-100	Apple, Papaya, Avocado, Plum, Ber and Pear
Very high	>100	Passion fruit, Sapota, Apple, Cherimoya

16. Based on Storage Life

Very perishable (0-4 weeks)	Perishable (4-8 weeks)	Semi- perishable (6-12 weeks)	Low perishable (>12 weeks)
Strawberry, Apricot, Berry fruit, Cherry, Fig and Loquat	Avocado, Grape Mandarin, Nectarin Passion Fruit, Peach and Plum	Orange	Apple, Grape Lemon and Pomegranate

Table 17A : All India Area, Production and Productivity of fruits, vegetables, flowers and aromatic plants over the Years 1991-92 to 2016-17 (Prov.)

Year	Fruits			Vegetables			Flowers & Aromatic		
	Area	Production	Productivity	Area	Production	Productivity	Area	Production	Productivity
1991-92	2874	28632	9.96	5593	58532	10.47	-	-	-
2001-02	4010	43001	10.72	6156	88622	14.40	106	535	5.05
2002-03	3788	45203	11.93	6092	84815	13.92	70	735	10.50
2003-04	4661	45942	9.86	6082	88334	14.52	101	580	5.74
2004-05	5049	50867	10.07	6744	101246	15.01	118	659	5.58
2005-06	5324	55356	10.40	7213	111399	15.44	129	654	5.07
2006-07	5554	59563	10.72	7581	114993	15.17	144	880	6.11
2007-08	5857	65587	11.20	7848	128449	16.37	166	868	5.23
2008-09	6101	68466	11.22	7981	129077	16.17	167	987	5.91
2009-10	6329	71516	11.30	7985	133738	16.75	183	1021	5.58
2010-11	6383	74878	11.73	8495	146554	17.25	191	1031	5.40
2011-12	6705	76424	11.40	8989	156325	17.39	760	2218	2.92
2012-13	6982	81285	11.64	9205	162187	17.62	790	2647	3.35
2013-14	7216	88977	12.33	9396	162897	17.34	748	3192	4.27
2014-15	6110	86602	14.17	9542	169478	17.76	908	3143	3.46
2015-16	6301	90183	14.31	10106	169064	16.73	912	3206	3.52
2016-17	6480	92846	14.33	10290	175008	17.01	943	3277	3.48

Areain'000Ha, Productionin'000MT, Productivity:MT/Hectare

Table 17 B : All India Area, Production and Productivity of Crops Plantation & Spices over the Years 1991-92 to 2016-17 (Prov.)

Year	Plantation Crops			Spices			Total		
	Area	Production	Productivity	Area	Production	Productivity	Area	Production	Productivity
1991-92	2298	7498	3.26	2005	1900	0.95	12770	96562	7.56
2001-02	2984	9697	3.25	3220	3765	1.17	16592	145785	8.79
2002-03	2984	9697	3.25	3220	3765	1.17	16270	144380	8.87
2003-04	3102	13161	4.24	5155	5113	0.99	19208	153302	7.98
2004-05	3147	9835	3.13	3150	4001	1.27	18445	166939	9.05
2005-06	3283	11263	3.43	2366	3705	1.57	18707	182816	9.77
2006-07	3207	12007	3.74	2448	3953	1.61	19389	191813	9.89
2007-08	3190	11300	3.54	2617	4357	1.66	20207	211235	10.45
2008-09	3217	11336	3.52	2629	4145	1.58	20662	214716	10.39
2009-10	3265	11928	3.65	2464	4016	1.63	20876	223089	10.69
2010-11	3306	12007	3.63	2940	5350	1.82	21825	240531	11.02
2011-12	3577	16359	4.57	3212	5951	1.85	23243	257277	11.07
2012-13	3641	16985	4.66	3076	5744	1.87	23694	268848	11.35
2013-14	3675	16301	4.44	3163	5908	1.87	24198	277352	11.46
2014-15	3534	15575	4.41	3317	6108	1.84	23410	280986	12.00
2015-16	3680	16658	4.53	3474	6988	2.01	24472	286188	11.69
2016-17	3677	16867	4.59	3535	7077	2.002	24925	295164	11.84

Areain'000Ha, Production in'000MT, Productivity:MT/Hectare

2

Use of Field Tools in Fruit Nursery

Major field operations for horticultural crops *viz.*, fruit nursery/seedling preparation, post hole digging for planting, intercultural operation, aeration, training and pruning, irrigation, plant protection, harvesting, handling, packaging transport are completed through field tools and implements. The cultivation of horticultural crops is predominantly dependent upon human labour, since commercial cultivation is only on a limited scale. So animal/power tiller or tractor-drawn mould board ploughs, disc ploughs, harrows, cultivators and rotavators are available for land preparation and development of new orchards. The fruit plants are generally perennial plants. So different types of sprayers and dusters are available from manually to power operated such as knapsack sprayers foot operated sprayers, power operated mist blowers and dusters or plant protection equipments used for application of insecticides, pesticides and herbicides in the fruit orchards. Some of important implements and sprayers are enlisted below:

S.No.	Common Name	Application	Implication
1.	**Spade**	It is useful to loosen the soil, prepare irrigation channels, collect the soil in heaps and facilitate filling up of soil, manure etc. in the baskets	
2.	**Khurpi**	For weeding and stirring the soil in the pots and beds	
3.	**Mattock**	It is a type of pick used for heavy work. One end of the blade is pointed as with a normal pickaxe, while the other is flattened like a chisel To loosen the soil in seedbeds and to break the clods	
4.	**Garden rakes**	Garden rake is a toothed rake used for collecting stones and bricks bits from the bed, scarifying the grass surface and gathering the fallen leaves	

5.	**Hand fork**	To loosen the soil in seedbeds and used for lifting and turning over soil in nursery
6.	**Spade**	Spade is useful for levelling of nursery bed make irrigation channel
7.	**Pruning shears**	It is very in useful garden as well as artificial necessary. Pruning, shears should not be very expensive but these should be made up of good steel. Similarly, these should make a smooth and clean cut with least injury to the fruit plant
8.	**Garden hand rake**	For removing, stubbles, small stones, levelling of nursery beds and formation of small beds.
9.	**Weeding fork**	It is widely used for soil loosening as well as in weeding. It consists of a long handle with a blade having teeth. It is drawn manually with the help of handle to collect the weeds
	Pickaxe	It is used for digging pits, making irrigation or drainage channels on compact soil and also for earthling up the rootstock
10.	**Sprayer**	It is useful for spraying insecticides, fungicides or herbicides. It is also used for foliar application of fertilizers
11.	**Hand Sprayer**	It is useful for spraying on small canopy plant to control the insect and pests
12.	**Sickle**	It is used for cutting standing grass, weeds, dry wood and harvesting crop

No.	Tool	Description	Image
13.	**Secateurs**	It is use for pruning, twigs, water suckers etc. of small plants. It is commonly both grafting, budding, and cutting. It has two sharp blades to give cuts on the stock and scion and the back end (flap) made up of plastic or brass used to lift or loosen the bark for inserting the bud	
14.	**Pocket Saw**	Use this small foldable saw to remove branches that are too thick for loppers (over one inch in diameter). It's portable and can be carried fairly easily in your pocket while you climb a ladder or a tree. (However, don't climb trees to prune unless you have the training and expertise to do so.) Most have tricut blades, which cut easier and faster than typical blades do. Gomboy and Gomtaru are two good-quality brands.	
15.	**Pruning knife**	It is commercially used for pruning of thicker branches and it has curved knife.	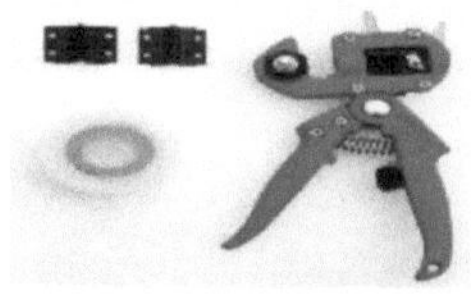
16.	**Pruning shear**	For cutting small size and diseases branches. It is commercially used in small and kitchen garden fruit plants	
17.	**Water cans**	These cans are used for irrigation of seeds in the nursery beds as well as for young plants in the pots. These are made up of galvanized iron sheet and sometimes these are also made up of plastic. These cans are fitted with a nozzle which is useful for equal distribution of water over the germinating seedlings	
18.	**Budding Knife**	A grafting knife, in general, has a straight 7.5 cm long blade and a strong long handle. It is use for are budding and grafting. The budding knife on the other hand may have a straight or a bit curved blade of shorter length. It is required for lifting or removing the patch of a bud from the bud wood.	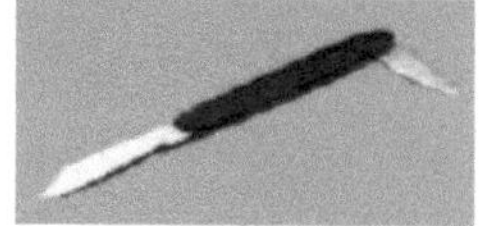

3

Nutrient Management in Fruit Crops

Soil

Soil is the upper most crust of earth surface which supports plant growth. It is defined as a three phase system in which plants grow. These phases are solid, liquid, and gas are essential. Solid part is frame which provides space for other two. It consists of minerals, clay minerals and organic matter. The soil is also a living system with millions of microbes that breakdown organic matter. The microbes are very essential and survive only when soil is well aerated and rich in organic matter and devoid of waterlogged conditions. Texture of soil depends on the size of solid particles and classified as gravel, coarse and fine sand, silt and clay. Soils are classified according to the relative distribution of these particles. Likewise, arrangement of these particles is referred as structure. Generally loamy soils and crumb structure are most growing preferred for fruit crops. According to level of organic matter, soils are classified as mineral soil or organic soil and soil having more than 20% organic matter is organic soil like peat and muck etc. Minerals and salts lend chemical properties to the soil like pH, alkalinity, sodicity, salinity and cation exchange capacity which influences the availability of nutrients in soil. Therefore, for making choice for soil, soil analysis in terms of following criteria isessential to decide on land capability. Maximum numbers of fruit orchards in India are depleted from micro and macro plant nutrients which interrupted the growth and development of the fruit plants and their quality. Therefore, like growers are not attain profit desired. To obtain marketable fruits, nutrients managements are highly beneficial to the fruit growers rainfed areas. Analysis of soil is useful to determine the level of the elements available in the orchards to get maximum yield.

Objectives

i. Importance of manures and fertilizers in fruit crops.

ii. Deficiency symptoms of nutrients in fruit crops.

All living organisms require certain elements for their survival. They are:

a. Basic elements: Carbon (C), Hydrogen (H) and Oxygen (O)

b. Macro elements: Nitrogen (N), Phosphorus (P), Potash (K), Calcium (Ca), Magnesium (Mg) and Sulphur (S)

c. Micro elements: Manganese (Mn), Molybdenum (Mo), Chlorine (Cl), Zinc (Zn), Boron (B), Copper (Cu) and Iron (Fe)

Macro elements: The nutrients that are required in relatively large quantity are termed as macro elements.

Micro elements: These elements that are required in relatively less quantity are termed as "micro nutrients". Besides some nutrients like Aluminium (Al), Cobalt (Co), Sodium (Na), Silica (Si) and Vanadium (V) are not considered necessary always because either their essentiality has been proved only in some plants or in certain metabolic processes that are not always necessary.

Organic manures

The waste products of plants and animals used as nutrients after decomposition referred to as "Organic Manures". These are the complex compounds from plants, animals and human residues that are used by plants as source of nutrient. They low in nutrient content and have long residual effects. Nutrients from manures are released only after decomposition of manure by micro organisms. Organic manures and leguminous green manures are most valuable from crop nutrition point of view. Besides, organic manures recycle the crop residues after decomposition. Organic resources reduce the mining of soil nutrients and improve physical properties of the soil by improving soil tilth, aeration, water-holding capacity and activity of the microorganisms. Manures are classified into two groups depending upon nutrient content they contain.

e.g. bulky manures:

a. Farmyard manure, compost, vermicompost, sewage and sludges.

b. Concentrated manures: Oil cakes, blood meal, meat meal and fish meal.

Farmyard manure (FYM)

Farmyard manure is blackish brown in colour, moist and sticky. Compost is mass of rotten organic matter made from farm waste. The decomposed mixture of dung and urine of farm animals along with litter and left over materials from roughages or fodder fed to cattle is farmyard manure.

Composting is a process in which both aerobic and anaerobic micro-organisms decompose organic matter at high temperature and low carbon-nitrogen ratio.

Table 1: Concentrated organic manures

A. Manures of plant origin:		**Nutrient content (%)**		
S. No.	**Plant origin**	**Nitrogen**	**Phosphors**	**Potash**
I.	Castor Cake	5.5 – 5.8	1.8 – 1.9	1.0 – 1.1
II.	Neem Cake	5.2 – 5.3	1.0 – 1.1	1.4 – 1.5
III.	Safflower cake	4.8 – 4.9	1.4 – 1.5	1.2 – 1.3
IV.	Coconut	3.0 – 3.2	1.8 –1.9	1.7 – 1.8
V.	Groundnut	7.0 – 7.2	1.5 – 1.6	1.3 – 1.4
VI.	Niger	4.7 – 4.8	1.8 – 1.9	1.8 – 1.9
VII.	Sesamum cake	6.2 – 6.3	2.0 – 2.1	1.2 – 1.3
B. Manures of animal origin:				
I.	Fish mean, fish manuring and fish guano	4.0 – 10.0	3.0 – 9.0	0.3 – 1.5
II.	Bone meal (Raw)	-	3.0 – 4.0	20.0 – 25.0
III.	Bone meal (Steamed)	-	1.0 – 2.0	25.0 – 30.0
IV.	Settled sludge (Dry)	2.0 – 2.5	1.0 – 1.2	0.4 – 0.5
V.	Night soil	1.2 – 1.3	0.8 – 1.0	0.4 – 0.5
VI.	Human urine	1.0 – 1.2	0.1 – 0.2	0.2 – 0.3
VII.	Cattle dung and urine mixed	0.60-0.80	0.15-0.40	0.45-056
VIII.	Horse dung and urine mixed	0.70-0.99	0.25-0.35	0.55-0.60
IX.	Sheep dung and urine mixed	-	3.0 – 4.0	20.0 – 25.0

Vermicompost

Vermicompost is granular, non-sticky and blackish brown in colour. The compost is prepared with the help of earthworms. It is a rich mixture of major and minor plant nutrients. It also increases total microbial population in the root zone and improves soil fertility.

Green manures

Green manures are the fast growing legume crops that are grown in the field prior to the cultivation of a main crop and is ploughing is done prior to incorporation into the soil for the purpose of restoring or increasing the organic matter content in the soil e.g. Dhaincha (*Sesbania canabina*), sunhemp (*Crotolaria juncea*), cowpea (*Vigna unguiculata*), horse gram (*Macrotyloma uniflorum*), berseem (*Trifolium alexandrium*), cluster bean (*Cyamopsis tetragonolaba)* and Lentil (*Lens culinaris*) etc.

Sources of manures

The sources of manures are as under:

- Cattle shed wastes that include dung, urine and slurry from biogas plants.
- Human habitation wastes e.g. night soil, human urine, town refuse, sewage, sludge and silage.

- Poultry litter, droppings of sheep and goat.
- Slaughter house wastes such as bone meal, meat meal, blood meal, horn and hoof meal, fish wastes etc.
- By products of agro industries like oil cakes, biogases and press mud, fruit and vegetable processing wastes.
- Crop wastes namely, sugarcane trash, stubbles and other related material and green manure crops.

Bio-fertilizers

Bio-fertilizers are the micro-organisms containing inputs which are capable of mobilizing nutrients from non-usable form to usable form through biological processes. They are less expensive, eco-friendly and sustainable. It improves plant growth and development by producing plant hormones. Some of the beneficial microorganisms are capable of fixing atmospheric nitrogen *viz.*, *Rhizobium, Azotobacter, Azospirillum* etc. On the other hand, some can increase the availability of phosphorus such as *Pseudomonas, Bacillus and Aspergillus* etc.

Benefit of human health

Residues of harmful pesticides in food or drinking water endanger both farmer's and consumer's health. Further health risks from antibiotics in meat, BSE infection (mad cow disease) and genetically modified organisms (GMO). In addition, this kind of agriculture is based on an excessive use of external inputs and consumes a lot of energy from non-renewable resources. It must be acknowledged that with the help of the green revolution technologies, crop yields increased tremendously, especially in the temperate zone. Several Southern countries also experienced the green revolution as a success story. However, the success of the green revolution in the south was unevenly spread; while, technology brought considerable yield increase in fertile river plains it rather failed on marginal soils, which constitute the major part of the land in the tropics. As the fertile lands usually belong to the progressive farmers, marginal farmers did not yield god from the new technologies. One reason for its failure on marginal lands is the low efficiency of fertilizer application on tropical soils: Unlike soils in temperate regions, many tropical soils do not retain chemical fertilizers well. The nutrients are easily leached out from the soil or evaporate as gas. The major part of the applied fertilizers may subsequently be lost. In countries where labour is comparatively cheap but inputs are expensive, expenses for agro-chemicals can make up a large proportion of the production costs. Frequently, these inputs are purchased on loans which are to be paid back when the harvest is sold. If yields are lower than expected (e.g. because soil fertility decreased) or crops entirely fail (e.g. due to attack of an unmanageable pest or disease), farmers still have to cover the costs of the agro-chemicals they used. Thus, indebtedness is a widespread problem among farmers in the South. As prices for agricultural products tend to decrease continuously while prices for inputs increase (e.g. due to reduced subsidies), it is becoming difficult for many farmers to earn sufficient income with conventional agriculture.

Benefits of organic manures

- Soil conservation and maintenance of soil fertility
- Less pollution of water (groundwater, rivers, lakes)
- Protection of wildlife (birds, frogs, insects etc.)
- Higher biodiversity, more diverse landscape
- Better treatment of farm animals
- Less utilization of non-renewable external inputs and energy
- Less pesticide residues in food
- No hormones and antibiotics in animal products
- Better product quality (taste, storage properties).

Inorganic fertilizers

Any natural or manufactured material, dry or liquid which are added to the soil in order to supply one or more plant nutrients other than lime or gypsum is known as "chemical fertilizer". These are industrially manufactured chemicals containing higher nutrient contents in comparison to organic manures and are soluble in form. In India five types of fertilizers are generally used in crop production.

i. Nitrogenous fertilizers
ii. Phosphatic fertilizers
iii. Potassic fertilizers
iv. Complex fertilizers
v. Mixed fertilizers

Compound fertilizers

Ammonium phosphate 20: 20 granulated fertilizer mixed fertilizers NPK (12:32:16) etc. Granular in form and brown or ash coloured Urea, single super phosphate and muriate of potash are important chemical fertilizers used in fruit production. Nutrients are lost from the soil through leaching, runoff, volatilization, fixation by soil or consumption by weeds etc.

Identification and nutrient content of chemical fertilizers

Handouts/material required/equipments & tools: Data sheet and pen to note down the observations, samples of organic manures and chemical fertilizers, petri dishes and beaker.

Table 2: Nutrient content in different fertilizers

Fertilizer	N (%)	P_2O_5(%)	K_2O (%)	Ca	S
Urea	46.0	-	-	-	-
Calcium Ammonium Nitrate	26.0	-	-	-	-
Ammonium Nitrate	34.0	-	-	-	-
Ammonium chloride	25.0	-	-	-	-
Calcium Nitrate	15.5	-	-	-	-
Ammonium sulphate	20.6	-	-	-	24.0
Single super phosphate	-	16.0	-	-	-
Rock phosphate	-	18.0	-	-	-
Bone meal	2	20	-	25	-
Potassium sulphate	-	-	52	-	16
Potassium chloride	-	-	60	-	-
Potassium Magnesium Sulphate	-	-	22	-	-

Liquid fertilizers	Nitrogen (%) by wt.	Water soluble P_2O_5 (%)	Water soluble K_2O (%) by wt.
Urea Ammonium Nitrate	32	-	-
Superphosphoric Acid	-	70	-
Ammonium Polyphosphate	10	34	-
Mix fertilizers	-	-	-
DAP	18	46	-
Potassium Nitrate	13	-	45
NPK (15:15:15)	15	15	15
NPK (10:26:26)	10	26	26
NPK (12:32:16)	12	32	16

Procedure/methodology

i. Spread the fertilizer sample on a piece of paper or in a petri dish. Note its colour. The colour may range from snow white to dark grey.

ii. Note the texture of the fertilizer which may vary from powder to globular granules. Some fertilizers have crystalline texture.

iii. Observe the hygroscopicity of fertilizer material. Hygroscopicity refers to the absorbance of water vapours from the atmosphere. The hygroscopic fertilizers usually form small to big lumps while the non-hygroscopic fertilizers maintain their original texture and do not form any lumps.

iv. Test the material for its solubility in water. Put a pinch of fertilizer in a beaker containing water. Stir it and carefully observe whether the fertilizer forms a solution or suspension over a time span of 5-15 minutes.

v. Hygroscopic fertilizers quickly get dissolve in water while others take a long time. Some may not dissolve at all and remain suspended in water.

Calculation of quantity of fertilizers based on nutrient requirement

Quantity of fertilizer

= quantity nutrient required x 100 ÷ nutrient present in fertilizer

Or

Q = N1 x100 ÷ N2

Where

Q = Quantity of fertilizer

N1 = Quantity nutrient required

N2 = Nutrient present in fertilizer

Or

Q = Quantity of nutrient required x factor

Calculation of quantity of complex fertilizers

First find out the quantity of fertilizer required for supplying of whole quantity of the major nutrients present in the particular fertilizer. Thereafter, calculate the amount of second nutrient supplied through the calculated quantity of fertilizer. Then subtract calculated amount of nutrient from the whole amount of second nutrient. Calculate the quantity of another source of fertilizers for balance nutrient quantity of second nutrient.

Calculation of quantity of chemicals for spray solution

$$V_1 = \frac{C_2 \times V_2}{C_1}$$

Where;

V_1 = quantity of chemical or commercial product

V_2 = volume of spray solution to be prepared

C_1 = nutrient content in chemical or commercial product

C_2 = concentration of spray solution

Methods of fertilizer application in fruit crops

Methods of fertilizer application

1. Broadcasting

It consists of spreading the fertilizers evenly over the entire field. This is applicable in full bearing or closely bearing orchards.

Disadvantages

Numerous elements like phosphorous and potash do not readily move in the soil. Therefore, surface application may not be available to the trees especially in drier tracks.

Leads to accumulation of potassium in surface soil beyond harmful levels causing injury to plants.

Surface application always stimulates weed growth.

2. Band placement

- Application of fertilizer on the sides of rows.
- Fertilizer in solid and liquid forms can be applied.

3. Ring placement

- Commonly followed in fruit trees.
- Fertilizers are applied in a ring encircling the trunk of the trees extending the entire canopy.
- It is more labour intensive and costly.

4. Foliar Application

- Fertilizers are applied in liquid form as foliar spray.
- They are easily absorbed by the leaves.

Fertilizers are applied in a very low concentration tolerable to the leaves.

Recommended when the nutrients are required in small quantity.

Foliar spraying can be used to apply nutrients like nitrogen or supplementary trace elements such as iron, magnesium, boron and calcium. Foliar applications are most effective during early growth.

The use of surfactants or spreaders, such as soap solutions, can enhance nutrient uptake through leaf tissue. To avoid leaf burn, nutrient sprays should not be applied during hot weather. Never apply pesticides and nutrients together.

The fertilizer may also be applied either as injections to the tree trunks or as dabs to cover the pruning wounds as is done in grapes. Application of concentrated fertilizers should not be applied too close to the trunks.

Each fertilizer application should be followed by copious irrigation.

5. Starter Solution

Liquid form of fertilizer application.

Seedlings and propagules are kept emerged up to their root system for varying duration in starter solution.

The starter solution is prepared either by dissolving concentrated fertilizer mixture at a concentration not exceeding 1%.

6. Fertigation

Application of fertilizers in irrigation water in either open or closed systems.

Nitrogen and sulphur are the principal nutrients applied.

Phosphorous fertigation is less common because of formation of precipitates takes place with high Ca and Mg containing water. Fertilizers that can be spread by irrigation include ammonium sulphate, ammonium nitrate, muriate of potash, calcium nitrate, ammonium phosphate, urea and borax. Do not use irrigation to apply phosphoric acid (a corrosive element), anhydrous ammonia (solutions will be lost to the air), or super phosphate and lime (cannot be dissolved in water).

7. Time of application

February to April the fruit trees have the highest requirements of nutrients for vegetation, flowering and fruit setting. To be effective, the nutrients must be readily available at a time when these are needed the most. The best time of applying N-fertilizers is about two weeks prior to the initiation of growth and flowering.

Table 3: Biochemical parameters as indicators of micronutrient deficiencies

Indicator	Nutrient
Nitrate reductase	Molybdenum
Catalase	Iron, Magnesium
Peroxidase	Iron, Magnesium
Phenol oxidase	Iron, Magnesium
Polyphenol oxidase	Copper
Ascorbic acid oxidase	Copper
Carbonic anhydrase	Zinc
Aldolase	Zinc
Ribonuclease	Zinc

Table 4: Soil reaction is measured in pH scale ranging from 0-14 and is meant to express the acidity or alkalinity of soil (soil suspension)

Soils are classified based on soil pH as follows:

S.No.	Soil condition	Soil pH rating
1	Strongly acidic	< 4.5
2	Moderately acidic	4.5-5.5
3	Slightly acidic	5.5-6.5
4	Neutral	6.5-7.5
5	Slightly alkaline	7.5-8.5
6	Moderately alkaline	8.5-9.5
7	Strongly alkaline	> 9.5

Table 5: Suitable in soil pH for crop growth

S.No.	Crop	Soil pH
1	Phalsa	6.0-7.5
2	Aonla	6.0-7.5
3	Karonda	6.0-7.5
4	Tarmind	6.0-7.5
5	Custard apple	5.5-6.5
6	Kinnow	6.0-7.5
7	Lasooda	5.5-7.0
8	Beal	5.5-7.0
9	Jackfruit	5.5-7.0
10	Ber	6.0-7.5
11	Guava	5.0-6.5
12	Fig	5.0-6.5
13	Mango	5.5-7.0
14	Citrus	5.5-7.5
15	Strawberry	5.5-6.5

Criteria for land capability class:

i. Slope and Erosion hazard

ii. Soil depth and Drainage

iii. Workability

iv. Stoniness and Rockiness

v. Water holding capacity

vi. Permeability

vii. Nutrient availability

viii. Fertility status

ix. Salinity, alkalinity and acidity hazards

Table 6: Alternatively tolerant or resistant crops can be chosen for different problems

S. No.	Salinity tolerant crops	Sodicity tolerant crops	Drought tolerant crops
1	Kair, Khirni, Woodapple, Date palm, Ber, Aonla, Fig and Sapota etc.	Ber, Tamarind, Woodapple, Date palm, Aonla, Karonda, Fig, Phalsa, Pomegranate, Guava and Bael	Ber, Aonla, Phalsa, Lasoda, Kair, Custard apple, Karonda, Pilu, Fig and Guava etc.
2	Fruits according to their tolerance to salinity:	Fruits according to their tolerance to Sodicity:	Fruits according to their drought tolerance:
a	**High salt tolerance:** Date palm, Ber and Aonla.	**High Sodicity tolerance :** Tamarind, Woodapple, and Date palm	**High drought tolerance** Kair, Pilu, Karonda, Ber and Lasoda
b	**Medium salt tolerance:** Pomegranate, Fig and Grape	**Medium Sodicity tolerance** Ber, Aonla, Karonda, Fig, Phalsa, and Bael.	**Medium drought tolerance**: Custard apple, Karonda and Fig.
c	**Low salt tolerance:** Orange, Lemon and Avocado	**Low Sodicity tolerance :** Guava and Pomegranate	**Low drought tolerance:** Guava

Soil analysis in terms of its physical and chemical attributes

- Bring the soil to its optimum potential by applying organic matter, chemical fertilizers, micronutrient and amendments depending on soil analysis report.
- Adoption of soil conservation technique like green manuring on regular basis.
- Use of improved water management techniques like drip irrigation and check basin or Furrows.
- Incorporation of large quantity of bulky organic matter each year.
- Creation of appropriate drainage around the plot.
- Scrapping of salts and reclamation of soil by application of gypsum, iron pyrites, press mud etc. on regular basis in case of salinity problem.
- Replenishment of nutrients harvested by the crop on regular basis by preparing a balance sheet for nutrients.
- Recycling of organic waste.

Table 7: Generalized symptoms of nutrients deficiency and excess in crops

Element / Status	Visual Symptoms
Nitrogen	Chlorosis begins in older leaves. Tips and margins of leaves commonly become yellow first, Stunted growth and Early crop maturity and low production.
Potassium	Symptoms first appear on recently matured leaves and become pronounced on older leaves. Dark green foliage with necrotic spots appearing on older foliage, typically at tips and margins and entire leaf blade becomes scorched or necrotic. Slow growth and susceptibility to wilting. Potassium is important for fruit taste, size and colour.
Phosphorus	1) Older leaves become dark green with purple coloration (due to anthocyanin development) in some species. 2) Stunted shoot growth and poor root development. 3) Delayed crop maturity.
Calcium	Reduced growth or even death of apical meristems, often leading to multiplebranching in tap root crops.Young foliage may be abnormal, chloritic or even 'burned' at tips. Softening of tissues and cell wall breakdown is common in fruits.
Boron (B) Deficiency	Abnormal development of the growing points (meristematic tissue) with apical growing points eventually becoming stunted and dying. Flowers and fruits will abort. For some grain and fruit crops, yield and quality is significantly reduced.
Excess	Leaf tips and margins will turn brown and die.
Chlorine (CI) Deficiency	Younger leaves will be chlorotic and plants will easily wilt. Excess premature yellowing of the lower leaves with burning of the leaf margins and tips. Leaf abscission will occur and plant will easily wilt.
Copper (Cu) Deficiency	Plant growth will be slow and plants stunted with distortion of the young leaves and death of the growing point.
Excess	Fe deficiency may be inducted with very slow growth. Roots may be stunted.
Iron (Fe) Deficiency	Interveinal chlorosis will occur on the emerging and young leaves with eventual bleaching of the new growth. When severe, the entire plant may be light green in colour.
Excess	Bronzing and tiny brown spots on the leaves.
Manganese (Mn) Deficiency	Interveinal chlorosis of young leaves while the leaves and plants remain generally green in colour. When severe, the plants will be stunted.
Excess	Older leaves will show brown spots surrounded by a chlorotic zone and circle.
Molybdenum (Mo) Deficiency	Symptoms will frequently appear similar to N deficiency. Older and middle leaves become chlorotic first, and in some instances, leaf margins are rolled and growth and flower formation and restricted.
Excess	Not of common occurrence.
Zinc (Zn) Deficiency	Upper leaves will show interveinal chlorosis with an eventual whiting of the affected leaves. Leaves may be small and distorted with a rosette form.
Excess	Fe deficiency will develop.

Table 8: General recommendation for amelioration of nutritional disorders

Deficiency	Soil application (kg ha^{-1})	Foliar Spray (%)
Iron	Application of ferrous Sulphate @ 50-100 kg ha^{-1}	Spraying of $FeSO_4$ 1 - 2%
Zinc	Application of zinc Sulphate @ 25 kg^{-1}	Spraying of 0.5% $ZnSO_4$
Manganese	Application of 25 kg Manganese Sulphate ha^{-1}	Spraying of 0.5% $MnSO_4$
Copper	Application of 5 to 10 kg copper Sulphate ha^{-1}	Spraying of 0.1% $CuSO_4$
Boron	Application of sodium tetraborate of borax at 10 kg ha^{-1}	Spraying of borax at 0.05%

Table 9: Leaf sampling technique and nutrient norms for fruits crops

Crop	Plant part, age, stage and position
Mango	Collect 4-5 month – old leaf from current season's growth from middle part

Symptoms of physiological disorders are confused with other nutritional deficiency or disease symptoms. Some other disorders of less importance nutrient have also been reported in many fruit crops, which are mention below.

Table 10: List of some minor physiological disorders/ diseases of fruit crops and their corrective measures

Name of disorder	Fruit	Responsible nutrient	Corrective measures
Exanthema	Citrus and Stone fruits	Copper deficiency	Spray of copper sulphate @ 0.1 % can control this disease
Anthracnose	Karonda	fungal disease *Colletotrichum inamdarii*	Spray copper sulphate (0.05%) or copper oxychloride (0.02%)
Corky core / spot necrosis	Apple, pear	Boron deficiency	Two sprays of boric acid (0.1%) at 15 days interval
Crook neck	Pineapple	Zn and Ca deficiency	Spray $ZnSO_4$ (1%) along with $CuSO_4$ (1%)
Die back	Pomegranate	Nutrients	
Fasciation	Pineapple	Environmental factors, tool Injury	Avoid tool injury
Fatio	Guava	Deficiency of Organic matter, nitrogen, zinc and boron	Supplement Soil with O.M, N, Zn and B
Heat spot / Kelsey spot	European plum	B deficiency and high temperature	Soil or foliar application of borax (500 ppm) in September and white washing are useful.
Isinasi	Pear	Ca and Mg deficiency	Application of Ca, Mg be increased
Leptonecrosis	Olive	- do -	- do -
Little leaf	Cashew	Zinc deficiency	Application of $ZnSO_4$ (0.5%) to the soil or through foliar means
Monkey face	Olive	B deficiency	Spray borax (0.1%)
Ring neck	Apricot	High K and low magnesium	Avoid excess of K and deficiency of Mg
Skin russeting	Persimmon	Excessive N and high relative humidity	Avoid excessive application of nitrogen
Snake heads	Walnut	B deficiency	Application of 1-2 kg borax/plant
Sunburn	Olive	B deficiency	- do -
Yellow leaf spot	Cashew	Molybdenum deficiency	Two sprays of ammonium molybdate (0.03%) during June and September
Yellow pulp	Banana	Excess of potassium	Maintain proper potassium and N level
Yellow/little	Walnut	Zn deficiency	Correct Zn deficiency by spraying 0.5% $ZnSO_4$
Ring neck	Apricot	High K and low magnesium	Avoid excess of K and deficiency of Mg

4

Cultivation of Minor Fruits

Minor fruit crops are considered important for their food, fibre, fodder, oil or medicinal properties, but have been reduced in value over time owing to particular supply and use constraints. These can include poor shelf life, unrecognized nutritional value, poor consumer awareness and reputational problems food or "poor people's food". The main products consumed by people are fruit cordial, jam, fruit drinks, chutney, candy, pickle and squash concentrate crop species. The entire country is mainly depend on handful of crop species neglecting large number of vegetables, fruits and other crop species with high nutritional and medicinal values which were grown traditionally in the past, resulting these valuable crop species are critically facing the state of speedy disappearing.

Major processed products which can be prepared from fruits

S.No.	Processed Product	Fruits
1	Jam	Jamun, Karonda, Aonla, Mulberry, Soursop, Tamarind and Wood apple
2	Jelly	Tamarind, Jamun and Karonda
3	Preserved	Ber, Aonla, Ker, Sangri, Karonda, Bael, Karonda, and Soursop
4	Candy	Aonla, Karonda and Tamarind
5	Canning	Ber, Aonla, jamun and Ker
6	Frozen Puree	Bael, Karonda, Ker, Phalsa Tamarind and Custard apple
7	Dehydration	Aonla, Karonda, Ker, Bael, Ber and Custard apple
8	Chutney	Karonda, Wood apple and Aonla
9	Sauce	Karonda, Tamarind, Wood apple, Pomegranate and Pickle
10	Wine	Mahua, Jujube, Ber, Indian fig and Karonda Jujube, Tamarind, Ker, Lasora and Gonda
11	Glazed Fruits	Tamarind, Annanas and Aonla
12	Confectionary	Amra, Aonla and Tamarind
13	Squash Juice/Syrup/ Beverage	Aonla, Ber, Bael, Jamun, Karonda, Phalsa, Mulberry, Pomegranate, Soursop, Wood apple and Tamarind

Importance of minor fruits

Indians are basically vegetarians, and to meet their nutritional requirement in terms of vitamins and minerals horticulture crops are to be grown in sufficient quantities to provide a bare minimum of 85 g of fruits per head per day with a population of above 120 cores. Continued increasing demand for horticultural produce provides tremendous scope for the growth of this industry. Good land is under pressure for staple food, industry, housing, roads and infrastructure due to population explosion. In India, large areas of marginal and wastelands, are not suitable for cultivation of cereals, pulses, and oil seeds crops etc., either due to poor quality soil or lack of water. Such lands are suitable for underutilized fruit crops such as jackfruit, ber, tamarind, custard apple, Indian gooseberry, phalsa and karnoda etc., as these crops can easily grow in poor soil condition and require less water for vegetative growth, flowering and fruiting. All these fruit crops are richest source of vitamins, minerals and antioxidants which are in good demand now-a-days but these crops are not very popular among the farmer's communities. Moreover all these fruits are good source of nutrients, vitamins, minerals, flavour, aroma, alkaloids, oleoresins, fibre, etc. Underutilized fruits are economic proposition as they give higher returns per unit area in arid and semi areas. It also enhances land value and creates better purchasing power for those who are engaged in this industry. Therefore, horticulture is important for health, wealth, hygiene and happiness. The biggest incentive for the farmer is money and underutilized fruits crops provide more returns in terms of per unit area production, export value, value addition comparedto agricultural crops. India is bestowed with a great variety of climatic and edaphic conditions varying from tropical, subtropical, temperate and within these humid, semi-arid, arid, frost free temperate etc. Soils viz., loam, alluvial, laterite, medium black rocky shallow, heavy black, sandy etc. are suitable for cultivation of these crops and thus a large number of crops can be accommodated with very high level of adaptability.

Necessity

After having achieved self sufficiency in food, nutritional security for the people of the country has become the point of consideration/priority.

The reasons for poor popularity of underutilized fruit crops are:

1. Lack of awareness about the economic benefits;
2. Non-availability of good quality planting materials;
3. Lack of technology to reduce the gestation period and enhance the fruit production;
4. Poor marketing network;
5. Lack of technology for value addition, through processing.

Most of these less known fruit trees establish through natural regeneration of seeds, grow slowly without any nutrition, start bearing fruits after a long period and produce fruits of inferior quality. Hence these species have remained neglected without any commercial importance. As some of these species are tolerant to harsh agro-climatic conditions, they have excellent potential for establishment on marginal and wastelands throughout the tropics. Fortunately during the last 1-2 decades, suitable technologies have also been developed to improve the productivity of these crops. However there is further need to set up field demonstrations to provide first hand exposure to the farmers for popularizing these species in the field.

Cultivation of minor fruit crops

Crop	**Soil and Climate**
Jackfruit	Jackfruit (*Artocarpus heterophyllus*) is a native of India. It requires rich and well drained sandy loam soil. Cold weather and frost are harmful to its cultivation. Hot desiccating winds in summer also adversely affect the growth of trees.
Aonla	Indian gooseberry (*Emblica officinalis*), locally known as *Anola* is used extensively for medicinal purposes. Deep sandy to sandy loams soils are best suited for aonla cultivation. However, good aonla orchards are found in clay loam and somewhat alkaline and saline soils too. The only requirement is assured irrigation during fruit development, i.e., May to September. A mature aonla tree can tolerate low as well as high temperature.
Beal	Bael is grown successfully on a wide range of soil types, *viz.* sandy loams, laterite, alluvial and calcareous soil even it is found growing well in stony, rocky and less fertile soils, but the better growth and higher yield is obtained in alluvial sandy loam soils with good drainage. However, owing to its hardy nature it can be grown in wide ranges of soil pH ranging from 5.0 to 8.0. Plant can tolerate high temperature also.
Ber	It can easily grow on a wide variety of soils-sandy, clayey, saline and alkali soils. The tree is remarkable in its ability to tolerate water-logging as well as drought. The tree can tolerate frost and a temperature range of 35-45°C.
Fig	It is grown successfully on a wide range of soil types, *viz.* sandy loams, and alluvial sand. The fig can be grown to height of 1525 m above sea level. In southern India variety marseilles is found growing on the hills and mountains. The tree can tolerate frost and a temperature range of 10-40°C.

Tamarind	It can grow well in any kind of soil provided there is sufficient moisture in the soil. If tamarind plants are grown in dry sandy and salty soils, they require frequent irrigation during the initial stage of their establishment, 10-15 days, especially in summer, till they are 5 to 6 years old. A compost of fibrous loam and sand suits the tamarind plants
Phalsa	It can be grown in tropical and subtropical region. It can withstand drought admirably. Since it is a very hardy tree, it can grow even in pockets of soil between crevices of barren rocks. Trees even grow on degraded and barren soil. The tree can tolerate frost and a temperature range of 40-42°C.
Jamun	Jamun can grow well under salinity and waterlogged conditions too. However, it is not economical to grow jamun on very heavy or light sandy soils. it can be grown under adverse soil and climatic conditions. It thrives well under both tropical and subtropical climate. It requires dry weather at the time of flowering and fruit setting.
Lemon	Lemon can be grown on various types of soils *viz*., clay, sandy loam and loam soil but deep loamy and alluvial soils are ideal for its cultivation. It performs well on sandy loam soil and attains good fruit colour, aroma and quality and plants can tolerate high temperature.
Karonda	It is a very hardy shrub and can be successfully grown in tropical and subtropical climates. Soil is not a limiting factor for its cultivation. It is grown successfully on a wide range of soil types, *viz*. sandy loams, laterite, alluvial sand, and calcareous soil even it is found growing well in stony, rocky and less fertile soils.
Pomegranate	It is grown in a wide range of soils; drought resistant and tolerant to salinity and alkalinity. Cool winter and dry summer are necessary for production of high quality fruits.
Pilu	Pilu plant can be grown on various types of soil *viz*. sandy, sandy loam, clayey loam, gravelly, shallow, calcareous and sand dunes with pH of 6.5 to 8.5. It can tolerate high temperature, plant can survive extreme drought and is very hardy. It can tolerate temperature within a range of -3 to +48°C; mean annual rainfall 180-1000 mm. It has been found growing both the in plains and in the hills up to an altitude of 900 m.
Mahua	It can grow on a wide variety of soils, but prefers sandy soils. It can withstand drought admirably and survive in rainfed areas where the normal annual rainfall is about 750 to 1875 mm.

Planting, manures and Fertilizers

Jackfruit An application of 75:60:50 g of NPK per year respectively up to 8 years and thereafter, the dose of 8th year taken as the constant dose for subsequent years, should be followed in jackfruit trees. It has been observed that the young fruits (0.5-1.0 kg) suffer from browning and mature fruits show the symptoms of developing spongy and corky tissues along with whitish pockets in the fruit mesocarp. This malady is believed to be caused due to lack of boron in the soil which can be controlled by spraying the trees with 1% solution of Borax at monthly intervals starting from January to May or by adding borax @ 250 g per tree along with fertilizer application. June-August is ideal time to planting.

Aonla Generally 1 to 2 year old plants should be given 10 to 15 kg of farm yard manure, 100 g nitrogen, 50 g phosphorus and 50 g of potash. This should be divided in three equal doses and applied 3 times in a year. The plant of 3 and more year should be given 20 kg farm yard manure, 500 g Nitrogen, 750 g phosphorus and 300 g of potassium. Aonla grow and perform well in all types of soil. The application of manure and fertilizers depends upon soil fertility, age of plant and production. Full dose of farmyard manure and P and half of N and K should be given in tree basin during January–February. The remaining half should be applied in August. In sodic soils, 100–500g of B and zinc sulphate should also be incorporated along with fertilizers as per tree age and vigour. Best time for planting is during July-August or February months.

Beal Apply 10kg farm yard manure, 50g N, 25g P and 50g K per plant to one year old plants. This dose should be increased every year in the same proportion up to the age of 10 years, after which the fixed dose should be applied each year. Half dose of N, full dose of P and half dose of K should be given after harvesting the fruits. Remaining half dose of N and K should be given in the last week of August. The best time of Bael planting is during raining season.

Ber Ber orchards are seldom manured. However, productivity of trees can be improved if manuring is done every year. The dose depends on fertility status of different locations. A dose of 75g N/tree gives highest yield, whereas 250g N and 250g P_2O_5 increase fruit yield. Application of K does not give response. In sodic soils, ber can be successfully planted even when the pH is over 8.5 and ESP is over 21 % by amending the soil of the pit prepared for planting by addition of gypsum. Begining of monsoon is the best time for planting.

Doses of fertilizers and manures for ber

Age	N (g/plant)	P (g/plant)	K (g/plant)	FYM
1	100	100	50	2
2	200	150	50	3
3	300	200	100	4
4	400	250	100	5
5	500	300	150	6

Fig

A general manure and fertilizer recommendation for fig is in table given below. For young plants fertilizers can be applied with the onset of monsoon and just after pruning for those which have commenced yielding. The annual requirement can be best divided into 2 applications, half after pruning and remaining 2 months later when the syconia are developing. Nitrogen is essential for rapid growth of foliage and development of syconia, fruit colour and maturation and K for yield and quality. Better fruit quality can be achieved if N and K are applied in the form of ammonium sulphate and sulphate of potash respectively. The best time for planting is the on set of the raining season.

Recommended dose of manures and fertilizers for fig

Age of plant (year)	Organic manure (Kg)		Inorganic manure (Kg)		
	FYM	Oil cakes	N	P	K
1-2	15.0	0.5	75	50	50
3-5	25.0	1.0-1.5	150	100	100
Above 5	40.0	2.0	300	200	200

Some soils may be deficient in micronutrients. General guidelines for correcting micronutrient deficiencies are mentioned in the table below. However, a grower should get the soil tested and consult the soil specialist for specific advice. Application of compost, which is done mostly in the beginning of monsoon also supplied micronutrients to some extent.

Micronutrients to be applied for correcting deficiencies

Micronutrient	Soil application	Foliar application
Zinc	30 kg $ZnS0_4$	3-4 sprays of 0.25% $ZnS0_4$ at 10-day interval.
Iron	-	3-4 sprays of 05% $FeSO_4$ at 10-day interval
Boron	12 kg Borax/ ha	-
Magnesium	50 kg $MgSO_4$/ha	2-3 sprays of 0.5% $MgSO_4$ at.10-day interval

Tamarind

Tamarind is a leguminous plant, tamarind may not require nitrogenous fertilizers but if the leaves which shed regularly are allowed to rot therein, it shall be beneficial in enriching the soil and improving its physical state. Phosphorus application at pit filling will be advantageous. The planting is done during the raining season.

Phalsa

Phalsa being a hardy crop is generally not given any fertilizers. Thus, an application of 15 kg F.Y.M after pruning followed by 125 g N/plant after sprouting is ideal for good yield. The fruit size was not affected by different levels of N, P and K, but high N and high P gave more fruit weight, as well as fruit quality. A fertilizer combination of 100 g N, 40 g P_2O_5 and 40 g K_2O gave maximum fruit yield. High levels of phosphorus supply increase sugar content in the fruit, while higher potassium suppresses sugar and promotes acidity. A combination of 200 g N, 25 g P and 50 g K increased fruit set, whereas 100 g N, 50 g P and 100 g K markedly reduced fruit drop in phalsa. Among the micro-nutrients, iron and zinc are found to influence berry size and juiciness. An application of 0.4 % $FeSO_4$ alone or in combination with zinc will improve the berry size. 4th leaf was considered to be suitable for nutrient diagnosis in *Grewia asiatica* monsoon season July-August in ideal time of planting.

Jamun

An annual dose of about 19 kg farmyard manure during the pre-bearing period and 75 kg per tree bearing fruits plant is considered. Normally, seedling jamun trees start bearing at the age of 8 to 10 years while grafted or budded trees come into bearing in 6 to 7 years. On very rich soils, the trees have a tendency to put on more vegetative growth with the result that fruiting is delayed. When the trees show such a tendency, they should not be supplied with any manure and fertilizer and irrigation. This helps in fruit bud formation, blossoming and in fruit setting. Sometimes this may not prove effective and even more drastic treatments such as ringing and root pruning may have to be resorted to. In pre-bearing period, 20-25 kg well-rotten cow dung manure or compost/ plant/year should be applied. For bearing trees, this dose is increased up to 50-60 kg/plant/year. The ideal time for giving the organic manures is month before flowering. Grown-up trees should be applied (5th year) 1 kg urea, 1.5 kg SSP and 500g MOP per plant per year.

Lemon

It is highly feeder fruit crop therefore proper nutrients management enhances the fruit yield and prevent the diseases incidence. The recommended manurial doses are very necessary. The first dose is applied in the month of February, second in June and third in September in a year. Planting can be done at the onset of the raining July-August.

FY M Kg/plant/ Nutrients (g)	**1yr**	**2yr**	**3yr**	**4yr**	**5yr**	**6yr n onwards**
FY M	20 kg 1 yr	25 kg 2 yr	30 kg 3 yr	35 kg 4 yr	40 kg 5 yr	40 kg 6 yr
N	100	200	300	400	500	500
P	50	100	150	200	200	200
K	25	50	75	200	200	200
$ZnSO_4$	25	25	50	50	50	50
$FeSO_4$	25	25	50	50	50	50
$MnSO_4$	25	25	50	50	50	50

Karonda

5 kg of FYM and 100 gm mixture of Nitrogen, Phosphorus and Potash should be applied in one year old plant. The four and more than 3 year old plants should given 15-20 kg of FYM and 400 g of mixture of NPK. The best time of fertilizer application is June-July after harvesting of fruits. Planting can be done at the onset of rainy season.

Pomegranate

It is a hardy fruit plant, growing successful by in L010 fertile soils. Application of 10 kg FYM and 75g ammonious sulphate to 5 years-old per tree. In general, application of 600-700 N gm 200-250g P_2O_5 and 200-250g K_2O per tree/year is optimum. Begining of monsoon is best time for planting of pomegranate.

Pilu

Pilu is a hardy fruit crop. But since the bearing is on new growth, it will respond to good fertilization. Thus, an application of 40 kg F.Y.M per plant/year. The best time for planting of Pilu is during monsoon season.

Mahua	Fully grown trees require 100 kg farmyard manure, 1 kg N, 0.5 kg p and 0.75 kg K. Farm yard manure should applied during July-August. Half dose of N and K and full dose of P should be applied a month before flowering and remaining half dose of N and K after fruit set. June-August is ideal time for planting.

Cultivars

Jackfruit	Jackfruit cultivars are classified into two types: 1) *Koozha chakka*, the fruits of which have small, fibrous, soft, mushy, but very sweet carpels 2) *Koozha pazham*, more important commercially, with crisp carpers of high quality known as *Varika*. The 'Singapore', or 'Ceylon', jack, a remarkably early bearer producing fruit in 18 months to 2-1/2 years from transplanting was introduced into India from Ceylon.'Safeda', 'Khaja', 'Bhusila', 'Bhadaiyan' and 'Handia' are excellent varieties' T Nagar Jack' as the best in quality and yield cultivar was introduced by The Fruit Experimental Station at Burliar.'Velipala', a local selection from the forest having large fruits with large carpers of superior quality. 'Black Gold' was selected in Queensland, Australia.'Cheena' and 'Cochin' was selected in Australia. 'Dang Rasimi' originated in Thailand. The tree is open, spreading and fast growing. 'Golden Nugget' was selected in Queensland, Australia. The tree is fast growing, with a distinctive dark green, rounded leaf.'Golden Pillow', or 'Mong Tong', was introduced to the Americas in the 1980s from Thailand.
Aonla	NA7, NA6, NA5, Anand-2, laxmi-52, CHES-1, Goma Ashwariya, Francies and Chakiya are important cultivars of Aonla.
Beal	NB-5, NB-9, NB-16, NB-17, Pant Urvashi, Pant Sujata and Goma Yashi.
Ber	Gaoma Kirti, Thar Sevika, Thar Bhubharaj, BS 75-1, Hybrid-1 and Narendra Selection-1 Mehrun, Tikadi, Ilayachi Tikadi, Banarasi Pebandi, Gola Gurgaon, Jhajjar selection, Umran, Darakhi 1, Darakhi 2, Guli and Vilayati
Fig	The varieties of fig predominantly cultivated in the state of Maharashtra are Poona fig, Dinkar, Dianna and Conardia. In Karnataka Bellary fig, Poona fig, Ganjam fig. Of late new varieties are also being introduced and the note worthy ones are Dienna and Dinkar. American fig varieties: Diredo, Faldders, Black Mission, Yovonne, Salebi, Tena, Renia, Nardinne Vienna, Gulban, Everem

Tamarind	There are no standard varieties of tamarind available as it is cultivated through seed. However, some selections are reported from Rahuri (Maharashtra) and regular plantations are being initiated in Maharashtra and Tamil Nadu. Also a great variation is available from sweet to sour type and these are utilized only after evaluation
Phalsa	CIAH-P-1, CIAH-P-2 and CIAH-P-3
Jamun	Goma Priyanka, Rahuri selection-1, Rahuri selection-2, AJG-85 and CISH–J-39
Lemon	Eureka, Lisbon, Lucknow Seedless, Nepali round and Pant lemon
Karonda	The cultivars of karonda are categories as per their colour of fruits *viz*., pink-white, greenish pink and reddish purple or on the basis of utilization. Pink varieties are white in colour at stage and turn to pink at maturity. Reddish, purple varieties are green at immature stage and turn to Reddish purple at maturity. The karonda varieties may be also classified in two categories i.e. pickle type varieties and table purpose varieties. Some varieties of Karonda have been developed during last two decades. Pant Manohar, Pant Sudarshan, Pant Suvarna are developed by GB Pant university of agriculture and technology. While Konkan bold, CHES K-II-7 and CHESK-35 are bold size and suitable? table purpose. K-1 K-2 & K-3 (Maharshatra), Selection No.3,12,13 & 16 (MPKV), Ruhuri, Pant Manohar Pant Suwarna, CHES K1, CHES-K2 and CHES-K-3 (Godhara)
Pomegranate	Important pomegranate varieties viz., Jyoti, Ganesh, Arakta, Rudhra, Bhagwa, Ruby, Alandi, Vadki, Dholka, Kandhari, Kabul, Muskati Red, Paper Shelled, Spanish Ruby, Ganesh (GB I), G 137, P 23, P 26, Mridula, , Jyoti, Ruby, IIHR Selection, and Co 1.
Pilu	There are no standard varieties of pilu available as it is cultivated through seed.
Mahua	Some selections have been made in ICAR research center and Agricultural University such as MH-10, MH-14, MH-35, MH-63, NM2, NM4, NM7, and NM8.

Propagation methods

Jackfruit	It is commercially propagated by cuttings, layering, air layering, budding and grafting and *in vitro* tissue culture. The success of the different vegetative propagation methods and the advantages

and disadvantages depend on the local climate, water availability and on suitable rootstocks.

Mound or Stool Layering

Stooling or mound layering is successful in jackfruit and treatment with IBA improves rooting of layered shoots.

Cuttings

Propagation by cutting is not so common. The cuttings with IBA (5,000 ppm) and P-hydroxybenzoic acid (200 ppm) at the time of planting was found effective in inducing rooting of semi-hardwood cutting.

Air Layering

It is another successful vegetative propagation method in jackfruit.

Grafting

Grafting is the most reliable method of propagation. Grafted jackfruit plant bears fruits in 2-3 years after planting and have a more spreading and open canopy than seedling trees.

Inarching: In south India, inarching of jackfruit is successful using *A. hirsute* or Rudrakshi as rootstock. Epicotyl grafting with mature, plump, terminal scion shoot on germinating jackfruit seedling of about 8-10 weeks by wedge method during April-May could be successful.

Epicotyl grafting: Epicotyl grafts attain saleable size within a year. The grafts become ready for planting in one or two years after grafting.

Aonla — It can be successfully propagated through patch / T-budding in north India during rainy season.

Beal — Bael is commonly grown from seed. Seedling plants show great variation in form, size, texture of rind, quantity and quality of pulp and number of seeds. The flavour ranges from disagreeable to pleasant. Therefore, superior types must be multiplied by asexual methods. A sexual methods *viz.,* T-budding and Patch budding in1-month-old shoots on to 2-year-old seedling bael rootstocks in the month of June-July gave better results.

Ber — Generally ber is propagated by T and patch budding methods. Rootstock seedlings are raised by sowing seed kernels extracted by breaking the stone (endocarp). These germinate in about one week. The seed stones can also be sown as such but take nearly

	one month to germinate. Seeds of any locally adapted and vigorous ber trees can be used for raising rootstocks in the field during July-August for in-situ budding or can be budded in the nursery beds. Budding is done during May-June on *Zizyphus mauritiana* rootstock.
Fig	Fig is commercially propagated by hardwood cuttings. Cuttings should be six to ten inches in length and approximately one-half to one inch in diameter. Place the cuttings in a warm, humid environment such as wrapping them in a moist paper towel and placed in a polyethylene bag for 10-14 days to encourage callus formation.
Tamarind	Seed and Grafting
Phalsa	It is commercially grown by seed. It is easiest and most common method to propagate the phalsa. Cuttings are difficult to root. Only 20% of semi-hardwood cuttings from spring flush, treated with 1,000 ppm NAA, and planted in July.
Jamun	Budding is practiced on one year old seedling stocks, having 10 to 14 mm thickness. Shield, patch and forkert methods of budding have proved very successful in jamun during rainy season.
Lemon	It is propagated through air layering and soft wood grafting during rainy season.
Karonda	Karnoda is propagated by sexual and asexual methods. Seed, stem cutting, air layering and budding are best for karonda.
Pomegranate	Layering and hard wood cutting are both successful methods of pomogranate propogation.
Pilu	It is commercially propagated by seed for natural regeneration occurs by seeds, layering and mostly by root suckers. Seeds are dispersed by birds, and often come up under other nurseshrubs such as *Capparis decidua* or *Tamarix spp*. Seedling trees get maturity after 7 to 9 years of planting.
Mahua	a) Soft wood grafting: It is practiced on one –year old seedling rootstock in the month of March –April may be used for better success by cleft method. b) Wedge grafting *in-situ*: It is practiced when seedling attain age of one year. It was observed that wedge grafting in-situ has better success in the month of March and April. c) Veneer grafting: It is done in the month of July and August.

Harvesting and yield

Jackfruit Jackfruit fruits are harvested for use as vegetables during early spring and summer until the seeds harden. Grafted plants starts bearing from the 5th to 6th years, while the tree reaches its peak bearing stage within 15th to 16th years after planting. Harvesting is done by cutting the stalks carrying the fruits. Normally, a tree bears a few to 250 fruits annually at this stage. The weight of the fruit varies widely depending on the type. Individual jackfruit may weigh upto 25 kg. Period of fruit development is February to June.

Aonla A grafted plant starts bearing after about 4-5 years of planting and harvest time in the month of December when they become greenish yellow from light green colour. A mature aonla tree of about 10 years will yield 50-70 kg. The normal weight of the fruit is 60-70 g and 1 kg contains about 15-25 fruits. A well maintained aonla tree yields up to an age of 70 years.A full grown grafted aonla tree with good bearing habit yields from 187 to 299 kg fruit per year and average fruit yield is 200 kg per grafted tree

Beal A grafted beal plant starts bearing after about 4-5 years and takes 10-12 months for ripening after fruit set. The average weight of the fruit is 200-400g and 1 kg contains about 2-3 fruits. Bael is climacteric fruit that can be ripened, off the tree, if harvested at proper maturity stage.

Ber The average yield during the prime bearing period (10-20 years) ranges 80-200 kg/tree. In dry areas, under rainfed conditions, 50-80 kg fruits/tree can be obtained. Fruits do not ripen after picking. Over-ripe fruits lose their eating quality and storage life. Therefore, fruits which are just mature and have shining yellow colour should be harvested.

Fig A good harvest may give 300 to 500 fruits per tree which depends on the size of the tree and method of training. Bearing in fig commences a year after planting, the life span of the tree being 35 years. The harvesting season varies with region and the yield depends on variety and cultivation practices. In North India, spring crop generally ripen in May, while in central and South India, the fig bears in July to September and February to May.

Tamarind Fruits are generally mature in February to March in which exocarp (outer most layer of the fruit) gets hardened and separated from

	the pulp. A fully developed tree can give production of 225 to 250 kg fruits per annum.
Phalsa	Phalsa fruits should be picked at right stage of maturity. Its fruits begin to ripe in hot summer. The best stage for harvesting phalsa fruit is at ripe stage during March-April in south and April-May in north India. An average fruit yield per bush from a well managed plant varies from 7-10 kg during one season.
Jamun	The fruit ripens in the month of June -July. The average yield of fruits from a full grown seedling tree is about 80 to 100 kg and from a grafted one 60 to 70 kg per year. These selected fruits are then carefully packed in wooden baskets and sent to the local markets.
Lemon	Lemon is potentially heavy bearers in subtropical or tropical climatic conditions. Normally about 70 to 200 fruits per year are expected from a five-year old or older tree in both crop.
Karonda	The fruits of karonda should be harvested when they attain proper maturity. About 7 to 8 kg fruits per is obtained both ripe as well as unripe fruits are harvested in subtropical climatic condition. Harvesting is normally done two to three times. Colour change is a good indicator of maturity of fruits.
Pomegranate	The fruits of pomegranate should be harvested when they attain proper maturity. Colour change is a good indicator of maturity of fruits. Normally about 100 to 130 fruits are expected in every year
Pilu	For high seed settings and seed oil content, harvesting is recommended 3 months after seed setting. This may be due to the utilization of food reserve in the cotyledons for the development of fruit pulp, and can be seen as the pulp content of fruit increases. Coppicing is advantageous for the tree's use as a fuel, and the branches are repeatedly cut to produce short stems that are harvested for toothbrushes. *S. persica* is grown in plantations or hedges. It is generally a slow-growing tree.
Mahua	Mahua yields both flowers and seeds. The ripe fruits shed from trees during June-July. TSS. Total and reducing sugars increased as fruit reached at maturity stage. The yield of flowers (dry) varies from 100-150 kg and kernel 60-80 kg.

Insects, Pests and Diseases

Jackfruit

Thirty-five species of insect pests have been recorded on jackfruit from India. However, the important of them are shoot and fruit borer (*Diaphania caesalis*), mealybugs (*Drosicha mangiferae* and *Nipaecoccus viridis*), bark eating caterpillar (*Indarbela tetraonis*), stem borer (*Batocera rufomaculata*), aphid (*Greenidia artocarpi*), scale insect (*Semelaspidus artocarpi*) and spiraling whitetly. Major insect pests infesting jackfruit are enlisted below. Shoot and fruit borer: *Glyphodes caesalis* (Walker), (Lepidoptera: Pyralidae) Spittle bugs: *Cosmocarta relata* Distant (Hemiptera: Cercopidae) Mealybug: *Drosicha mangiferae* Stebbins (Hemiptera: Margarodidae) Bud weevil: *Ochyromera artocarpi* Marshall (Coleoptera: Curculionidae) Bark eating caterpillar: *Indarbela tetraonis* (Moore) (Lepidoptera: Cossidae) Aphid: *Greenidea artocarpi* (Westwood) (Hemiptera: Aphididae) Leaf webber: *Glyphodes bivitralis* Gueneé (Lepidoptera: Crambidae).

Shoot and fruit borer

The adult moth is about half-an inch long with black edge bands and pale yellow patches on wings. The moth lays eggs on leaf, shoot and flower bud. The first instar larva is very tiny. The full-grown caterpillar is pinkish, each segment banded with numerous black flattened horny warts from which arise single short bristly hairs. Head and prothoracic shield become yellow. Pupation takes place in a silken cocoon made by the caterpillar inside the fruit tunnel several inches long, twisted dried of leaf or on the surface of nearest two more fruits. Pupa is red brown in colour and pupal period is about a week. Adults are whitish-brown moths having wings with grayish elliptical patterns and a marginal series of black specks; wing expanse is 26 to 30 mm, females being slightly bigger than males.

Nature and extent of damage

In India the pest is active from May to October. Eggs are laid on tender shoots and flower buds. On hatching the caterpillars bore into tender shoots, flowering buds and developing fruits and tunnel through the same. Severe infestation on fruits may induce drop. Early infestation of jackfruit borer results in deformation of fruits and sometimes dropping off the immature fruits. This jackfruit borer also attacks in nursery. Larvae bore into the tips of jackfruit saplings and proceeds towards the base by making tunnel. As a result of attack, the affected parts wilt and dry resulting lateral branching of sapling. For monitoring the incidence of insect pests following method is to be employed.

Mealybug: Count and record the number of both nymphs and adults on five randomly selected leaves per tree.

Leaf webber: Count the number of webs formed in each direction, thus covering the whole tree.

Scale insects: Number of scale infested shoots per five tender shoots from each of the four directions of the selected tree should be counted.

Defoliator/ borers: Count the number of young and grown up larvae on each plant and record.

Management Strategies

- Keep soils covered year-round with living vegetation and/or tree residue.
- Add organic matter in the form of farm yard manure (FYM), vermicompost, tree residue which enhance below ground biodiversity of beneficial microbes and insects. Application of balanced dose of nutrients using biofertilizers based on soil test report.
- Application of biofertilizers with special focus on mycorrhiza and plant growth promoting rhizobacteria (PGPR).
- Natural enemies play a very significant role in control of foliar insect pests. Natural enemy diversity contributes significantly to management of insect pests.

Natural enemies may require

- Food in the form of pollen and nectar.
- Shelter, overwintering sites and moderate microclimate, etc.
- Alternate hosts when primary hosts are not present. In order to attract natural enemies following activities should be practiced:
- Raise the flowering plants/compatible cash trees along the orchard border by arranging shorter plants towards main tree

and taller plants towards the border to attract natural enemies as well as to avoid immigrating pest population.

- Grow flowering plants on the internal bunds inside the orchard.
- Do not uproot weed plants those are growing naturally such as *Tridax procumbens*, *Ageratum* sp, *Alternanthera* sp etc. which act as nectar source for natural enemies,
- Do not apply broad spectrum chemical pesticides, when the P: D ratio is favourable. The plant compensation ability should also be considered before applying chemical pesticides.
- Reduce tillage intensity so that hibernating natural enemies can be saved.
- Select and plant appropriate companion plants which could be trap trees and pest repellent trees. The trap trees and pest repellent trees will also recruit natural enemies as their flowers provide nectar and the plants provide suitable microclimate.
- Due to enhancement of biodiversity by the flowering plants, parasitoids and predators (natural enemies) number also will increase due to availability of nectar, pollen and insects etc. The major predators are a wide variety of spiders, ladybird beetles, long horned grasshoppers, *Chrysoperla*, earwigs, etc.

Management strategies

Shoot and fruit borer

- Attacked shoots should be clipped off and destroyed.
- Clean hole and pour kerosene/petrol/crude oil or formalin into the stem borer hole and subsequently close entrance of the tunnel by plugging with cotton wool and paste the mud.
- Use light trap@1/acre
- To protect them from egg laying, fruit may be covered with polythene bags and the affected parts removed and destroyed.
- Spraying neem oil may be recommended.

Spittle bugs

- Keep orchard clean and healthy.
- Cut dried branches.
- Light, accessible spittlebug infestations can be removed by hand or by a strong water spray.
- Pipunculid fly, *Verrallia virginica* caused 50-60% parasitism of adult spittlebugs.

Mealybug

- Flooding of orchard with water in the month of October kill the eggs. Ploughing of orchard in November.
- Raking of soil around tree trunk to expose the eggs to natural enemies and sun, removal of weeds Fastening of alkathene sheet (400 gauge)/grease band of 25 cm wide afterwards mud plastering of trunk at 30 cm above the ground in the middle of December. In July-August destruction of infested fallen leaves is must.

Scales

- Raking of soil around tree trunk to expose the eggs to natural enemies and sun, removal of weeds and releasing 10-15 grubs.
- Releasing 10-15 grubs of cocinellid predator, *C. montrozieri* per tree.

Bud weevil

- Remove the infested shoots, flower buds and fruits to check infestation.

Bark eating caterpillar

- Remove and destroy dead and severely affected branches of the tree.
- Remove alternate host, silk cotton and other hosts.

Aphid

- Collect and destroy the damaged plant parts along with nymphs and adults.
- Release coccinellid predators.

Pink waxy scale

- Prune heavily infested plant parts to open the tree canopy and destroy them immediately.
- Prune infested parts (branches and twigs) preferably during summer. These should be placed in a pit constructed on one corner of the orchard. Allow branches and twigs to dry until the parasites escape.
- Burn the remaining debris.
- Removal of attendant ants may permit natural enemies to control the insect.

Leaf Webber

- Mechanical clipping and burning of affected shoots.
- Biological control: Release of pupal parasitoid, *Tetrastichus howardi* @20,000 / ac.
- Release of egg parasitoid, *Trichogramma chilonis* @2cc/ac.

Thrips

- Spraying strong jet of water to dislodge and wash out the pest Castor capsule borer.
- Prune heavily infested plant parts to open the tree canopy and destroy' them immediately.
- The natural enemies, *Hexamermis* spp and *Apanteles taragammae* have been found to be potential bio-control agents of the pest.

Aonla

Insect Pests

Bark-eating caterpillar: [*Indarbela tetraonis* (Moore)]

Symptoms

The pest is identified by the presence of irregular tunnels and patches covered with silken-web consisting of excreta and chewed up wood particles, on the shoots, branches, and trunk. On removal of the bark or infested silken web, larvae can be seen hiding beneath. Shelter holes may be seen particularly at the joints of shoots and branches. The young shoots dry and die, giving a sickly look to the tree.

Management strategies

- Keep the orchards clean and healthy to prevent the infestation of this pest.
- Detect early infestation by periodically looking out for dried or drying young shoots.
- Kill the caterpillars mechanically by inserting the iron spike in shelter holes made by these borers at early stage of infestation. In case of severe infestation, remove webs and insert swab of cotton wool soaked in 0.025% dichlorvos or inject water emulsion of chlorpyriphos (0.05%) or kerosene oil and plug the holes
- The larvae are parasitized by entomogenous fungus *Beauveria bassiana* in nature. It can be used as a potential bio-control agent.

Shoot gall maker: [*Betousa stylophora* (Swinhoe)]

Symptoms As a result of attack by this pest terminal shoots swell that increases in size with the passage of time. Full sized galls can be seen in the month of October-November.

Management Strategies

- Overcrowding of branches should be discouraged. Galled shoots should be pruned and destroyed along with the pest after harvest.
- In case of regular occurrence of this pest, spray chlorpyriphos (0.05%) in the beginning of the season. It may be repeated at fortnightly intervals, if needed.

Pomegranate butterfly: [*Deudorix (Virachola) isocrates* (Fabr.)]

Symptoms

Affected fruits are generally deformed at the point of entry of larvae. Frass may be seen exuding out of the borer hole. Such fruits weaken, rot and fell down before maturation. In case of severe attack, it may cause considerable loss. Attack of this insect occurs during September-October, coinciding well with the fruiting season. The violet brown female butterfly lays shining white eggs, singly on young fruits. The larva of this pest bores the fruit and feeds on seeds, making this portion hollow from inside.

Management strategies

- Cultivation of pomegranate and guava should be discouraged close to aonla plantation as these are major host plants of this pest.
- Infested fruits should be identified, collected and destroyed to prevent further spread of infestation.
- Spray spinosad 0.25 ml/l or carbaryl (2g/l) at pea size stage of aonla fruits. The spray may be repeated after two weeks, depending upon the intensity of attack.
- Release of *Trichogramma chilonis* @ 2.5 lakhs/ha four times at 10 days interval.

Mealy bug: [*Nipaecoccus viridis* (Newstead), N. *vestator* (Newstead)]

Symptoms

The attacked new shoots are found bending and twisting with yellowing of leaves. In case of severe infestation, twigs become leafless and dry. Excessive excretion of honeydew is noticed. Flowers dry up and drop.

Management strategies

- Clean cultivation and maintenance of health and vigour of the tree.
- Prune affected parts and destroy them at early stages of infestation.
- In case of severe infestation spray spinosad (0.25 ml/l).

Aonla aphids: [*Cerciaphis emblica* (Patel & Kulkarny), *Schoutedonia emblica* (Patel & Kulkarny) and S*etaphis bougainvillea.* (Thunberg)]

Symptoms

The infested leaves turn yellow and dry up. Infested shoots appear bended and twisted at the growing points. Presence of ants also indicates the infestation of aphids. The new shoots are infested at growing points. The nymphs and adult females suck the sap. Heavy attack affects the growth and vigour of the tree, ultimately affecting the flowering and fruiting.

Management strategies

- Clipping off and destruction of affected leaf and shoot.
- Spray dimethoate (0.06%) or spinosad (0.25 ml/l).

Leaf rollers: [*Garcillaria acidula* and *Tonica (Psorosticha) ziziphy* Stainton]

Symptoms

The infestation may be identified by webbing of leaves, their withering and dropping. The adults of these insects are miniature moths. Larvae of these moths bind the leaves together and feed therein. In case of heavy incidence leaves dry and drop leading to drying of twig also.

Management strategies

- Avoid overcrowding of branches and maintain sanitation in the orchard.
- Rolled leaves may be clipped off and destroyed along with the larvae in the beginning of infestation.
- In case of heavy incidence, spray carbaryl (0.2%) or chlorpyriphos (0.04%).

Stone borer: [*Curculio* sp.]

Symptoms

Externally the ovipositional site appears as small brown patch on the fruit. The infested fruits may also be identified by the presence of exit hole. Tiny weevil emerges in the month of June with the onset of rains. Emergence continues in July-August also, coinciding well with the fruiting season of aonla. The eggs are laid by excavating a cavity below the epicarp. Fruits of 1.5 to 2 cm in diameter are preferred for oviposition. Larva after hatching travels through the mesocarp, reaches to stone, enters inside it and feed there on sees, destroying them completely.

Management strategies

- Deep ploughing of the orchards after harvesting exposes the diapausing larvae and is effective in bringing down the pest population.
- First spray of carbaryl (0.2%) or chlorpyriphos (0.05%) at pea size of fruit. Second spray may be done at fortnightly interval with changed insecticide, if needed.

Fruit midge: [*Clinodiplosis* sp.]

Symptoms

In the beginning, small grey black spots appear at the site of infestation. It darkens to brown at later stage. The emergence hole is minute but visible easily. The affected parts of fruit rot because of the damage done by the midge larvae and also fruits become susceptible to secondary infection by different pathogens. Desi varieties of aonla are more susceptible. Incidence of this pest occurs in the fruiting season in aonla from September to January. Adults are seen on wings in the months of September-October and are miniature flies. The eggs are laid inside the fruits. Larvae after hatching feed on the content of the fruits. Immature larvae are creamish-white, while mature are pinkish-orange in colour. The mature larva is about 3-3.5 mm in length and 1 mm in width. Fully matured larvae jump out of the fruits, drop down in the soil and pupate there. Full life cycle is completed in 20 to 28 days, with the last larval stage diapausing within the soil.

Management strategies

- Deep ploughing of the orchards after harvesting exposes the diapausing larvae and is effective in bringing down the pest population.
- Spray of carbaryl (0.2%) or chlorpyriphos (0.05%) at the beginning of the fruiting.

Ber

Though as many as 130 species of insect pests have been recorded in India, only few species have attained the pest status and cause substantial economic damage to ber. Out of these, three insects viz., ber fruit borer, *Meridarchis scyrodes* Meyrick and two species of ber fruit flies, *Carpomyia vesuviana* Costa and *Dacus correctus* (Bezzi) were recorded as major pests on this crop with infestation varying from high to very high degree, whereas five insects (Grape mealy bug, *Maconellicoccus hirsutus* (Green); Ber mealy bug, *Perissopneumon tamarindus* (Green); Ber fruit weevil, *Aubeus himalayanus* Voss; Castor semilooper, *Achaea janata* Linn. and Snail, *Cryptozona semirugata* (Beck)) were recorded as moderate pests. As many as nine insect pests viz., Cow bug, *Tricentrus bicolor*; Thrips, *Scirtothrips dorsalis* (Hood); Longicorn beetle, *Celeosterna scabrator* Fabricius; Gray weevil, *Myllocerus discolor* (Boheman); Spiny beetle, *Platypriya andrewesi* Weise; Hairy caterpillar, *Thiacidas postica* Walker; Tassar silk moth, *Anthearea paphia* Linnaeus; Tobacco caterpillar, *Spodoptera litura* (Fabricius) and Eriophyid mite, *Eriophyes cernus* Massee were recorded as minor pests on this crop. Five insect pests including Green striped leaf hopper, *Eurybrachys tomentesa* Fab.; Jassid, *Amrasca biguttula biguttula* (Ishida); Spittle bug, *Machaerota planitiae* Distant; Lac insect, *Laccifer lacca* Kerr. and Ber butterfly, *Tarucus theophrastus* (Fabricius) were recorded as negligible pests.

Damaging Symptoms

Ber butter Fly

Larvae feed on sprouting tender shoots, leaves and flower buds. Infested leaves gives whitish look due to chlorophyll feeding. Finally the leaves remain with long streaks.

Leaf webber

The newly hatched caterpillar attacks unopened and partially opened leaves and leaf folding with silken threads. The larvae consumes green matter by scrapping, leaving behind the papery epidermis. In severe state, the tree gives unhealthy appearance and stunted growth on growing point.

Fruit fly

Maggots start feeding on pulp and the infested fruits packed with excreta of maggot. Under severe conditions, fruits drop off.

Bark eating caterpillars

Presence of webs at angles, weakening and cleavage of branches at fruit development stages are the typical symptoms of this pest attack.

Grey weevil

The damaged leaves have serrated margin and webbed leaves.

Termites

Galleries of termites and their mounds may be noticed.

Mites

Both nymphs and adults suck the cell sap, thus devitalizing the plant and reducing fruit yields. Leaves give a sickly appearance.

Management Strategies

- Repeated ploughing around the tree destroys different stages of the insects in the soil.
- Apply Neem Seed Cake after ploughing @ 4 kg / tree.
- Repeated hoeing / weeding around the tree basin.
- The mite affected leaves and twigs should be cut and burnt in the month of May and October.
- Fix the reflective ribbons at ten feet distances in the orchard to prevent parakeet.
- A number of effective parasites, predators and pathogens can be used against pests of ber. Eg. Spiders, *Sumnius renardi*, Coccinellids, *Chrysoperla lacciperda* which could be conserved by using various conservation methods.
- Neem Seed Kernel Extract @ 5% helps in reducing the pest population.
- Dicofol 18.5% or Kelthane (3.0 ml / l of water) should be applied twice against Ber mite at 10 days interval.
- Spray Carbaryl (2 g/L) for control of fruit borer and leaf miner.

Tamarind

Tamarindus indicus, native to the dry savannah of tropical Africa is an important tree species widely grown and cultivated throughout India. Tamarind fruits are used and relished as an important ingredient in 'curries' 'chutnies' and various other preparations. Insect pests involved in large scale damage to the various parts of the tree, are summarised. A number of sapsuckers, mealy bugs, scale insects, aphids suck the sap of the tender shoots and leaflets. Some species of caterpillars and beetles cause damage to the foliage, flowers, fruits and seeds. Control measures of various pests are also given.

Tamarind/Peanut bruchid: *Caryedon serratus*

Symptoms of damage

- Grub causes the damage.
- Circular hole on fruits and seeds of tamarind both in tree and storage.

Identification of the pest

Egg : Laid singly, glued to the surface of the pod (in fields) or on grains (in stores). Fresh eggs are translucent, orange cream in colour, changing to greyish white with age.

Grub : Fleshy, curved creamy white in colour with black mouth parts.

Pupa - Pupation takes place in a pupal cell prepared beneath the seed coat.

Adult

- Brownish grey beetle with characteristic elevated ivory like spots near the middle of the dorsal side.
- It is small, short, and active with long conspicuous serrate antenna.
- Elytra do not cover the abdomen completely, which is called as pygidium.

Management Strategies

Phalsa

The following pests can be monitored by the procedure mentioned below. Aphids, mealybug, phalsa bug, psylla: Count and record the number of both nymphs and adults on five randomly selected leaves per plant. Defoliator/ borers: Count the number of young and grown up larvae on each plant and record.

Management Strategies Aphids

- Judicious use of nitrogenous fertilizers.
- Spray azadirachtin 5% W/W neem extract concentrate @ 80 g in 160 l of water/acre. Release 1st instar larvae of green lacewing @ 4000/acre.

Phalsa bug, Bark eating caterpillar and hairy caterpillar

- Keep orchard clean and healthy.
- Bark eating caterpillar infested galleries should be removed.
- Collect and destroy gregariously feeding larvae of hairy caterpillar.
- Keep the orchard free from weeds to avoid pest infestation.
- Release of egg parasitoid *Trichogramma chilonis* @ 20,000/ acre
- Use of 5% NSKE.

Brown and phalsa beetle

- Keep orchard clean and healthy.

Mealybug

- Remove weeds like *Clerodendrum inflortunatum* and grasses by ploughing during June-July.
- Plough orchards during summer to expose the eggs to natural enemies.

Integrated Management Strategies to be followed for phalsa pests

Cultural control

- Integrated strategy is essential to curtail the pest attack from nursery to field level.
- Proper pruning of shrubs during month of December to February and racking of soil around the plant will reduce the residual pest population especially chafer beetles and fruit fly pupa.
- Manual removal of leaf miner infested leaves, webs of bark eating caterpillar and caterpillar larvae would encourage the pest free condition.
- The leaf miner affected leaf should be pruned in nursery as well as main field.

- Collection and destruction of hairy caterpillar larvae also would be effective in controlling this pest.
- Selection of healthy and leaf minor free seedlings for planting, periodical cleaning of orchards, racking of soil around the shrub and timely pruning is necessary to reduce the pest attack.

Mechanical control

- Installation of fruit fly trap catches using methyl eugenol + malathion for male annihilation.
- Light trap can be used to catch the adult chafer beetle during night and to monitor the moth's activity.

Botanical pesticides

- Application of neem seed kernel extracts 4% (NSKE) will effectively reduce the pest infestation and also help to conserve the existing natural enemies.

Biological control

- Natural predation of coccinellid predator, *Coccinella sexmaculata* and *Scymnus* sp has been reported on aphid. Predator coccinellid, *Cryptolaemus mountrouzieri* and lycaenid predator, *Spalgis epius* found to occur on pink mealy bug

Chemical control

- Foliar spray of dimethoate (0.03%) during sprout emergence and imidacloprid 0.008% or thiomethoxam 0.008% reported had significantly high nymphal reduction of psylla and leaf minor.
- Application of carbaryl 2.5g/lit of water would be effective against leaf feeders.
- Application of petrol or kerosene or dichlorvos on hole and plugging with cotton or mud immediate after pruning for bark eating caterpillar will reduce the damage.
- Bait spray consist of malathion and jaggery will effectively reduce the fruit fly infestation.

Citrus — Citrus psyllid, black aphid, fruit sucking moths, mealy bug, leaf miner, citrus scale, etc. are the major insect pests attacking citrus trees.

Management strategies

Collect and destroy the damaged plant partsEncourage activity of parasitoids, *Encarsia sp.*, *Eretomocerus serius* and *Chlysoperla sp.* and predators like Syrphids and Chrysopids

- Field release of Australian lady bird beetle *Cryptoleamus montrouizeri* 10 per tree
- Use sticky trap (5cm length) on fruit bearing shoots.
- Apply mixture of manure compost tea, molasses, citrus oil.
- Control ants and dust which can give the scale a competitive advantage.
- Field release of vedalia and Australian ladybugs.
- Destroy the weed host *Tinospora cardifolia* and *Coccules pendules* of fruit sucking moth
- Bag the fruit with polythene bag (500 gauge)
- Apply smoke to prevent adult moth
- Trap crop – growing tomato crop in orchards to attract the adult moth
- Poison baiting with dilute suspension of fermented molasses and malathion 0.05% (50 EC at 1ml/lit)
- Use light trap or food lure to attract moths.
- Spray dormant oil in late winter before spring.
- Spray horticultural oil, if needed, year round.

Karonda

Diseases

Anthracnose and bacterial canker are the common diseases of karondai.

Anthracnose

This fungal disease is caused by *Colletotrichum inamdarii*. This disease is widespread in U.P. and can be seen in the form of black grey spots on the leaves. Spray of copper sulphate @ 0.1 % can control this disease.

Bacterial canker

This is a bacterial disease and caused by *Xanthomonas carissa*. Symptoms first appear as spots on the lower surface of the leaf. These are small, round and water soaked which turn dark brown.

A yellowish green halo may surround the spots which causes necrosis. This disease can be checked by removing diseased leaves followed by a spray of Phytomycin (200 ppm).

Pomegranate **Pests:** Approximately all varieties of Pomegranate are found to be susceptible to the anar butterfly (*Virachola isocrates*) which is the most important pest in Sri Lanka. The caterpillar of the butterfly enters the fruits and feed on it.

Diseases: In Sri Lanka, there are several fungus diseases which are economically important in Pomegranate cultivation and are detailed below.

Fruit roti. Anthracnose fruit rot - Caused by *Sphaceloma punicae*ii. Aspergillious fruit rot - Caused by *Aspergillus spp.*iii. Penicillum fruit rot - Caused by *Penicillium spp.*

Leaf spots

Causal organisms are Colletotrichum spp., Phomopsis spp., Cercospora punicae.

Die back

Caused by either *Pleuroplaconema* spp. or *Centhosphora phyllostica* Fruit cracking Cause is not fully understood. However it is suspected to be a physiological disorder.

Pilu Extreme agro climatic conditions in arid zone not only generate greater biodiversity in plants but also provide congenial ground for perpetuation and manifestation of varied pests and diseases caused by different groups of pathogens. *Salvadora persica*is susceptible to many diseases and pests. Both seedlings and mature plants are affected by viral diseases, especially crown gall. Fungi such as *Cercospora udaipurensis, Phytoplasma sps., Placosoma salvadorae* and *Septogloeum salvadorae* also damage the leaves of *S. persica.*

At the same time, *Thomasiniana salvadorae*, induced gouty galls which were also noticed on the mid vein of *Salvadora persica.* Moreover, termite amitemes belli (*Desneuse*) eat heart wood and sapwood of Salvadorasps.

Mahua The most important Fungi diseases like *Aspergillus flavus, A. niger, Penicillium* sp. *and Stathmopoda basiplectra* causes deterioration in seeds and fruits. Leaf spots, leaf blight and leaf rust diseases are commonly found in nursery stage as well as in maturity stage of plants.

Parasites: *Dendrophthoe Falcate* (L.f.) is serious parasite for attacking the growth and productivity of Mahua. There is considerable reduction in the formation of inflorescence and the flower size of the plant. The parasite simultaneously weakens the plants and on ageing the plant dies. The parasite can be controlled by applying certain weedicides like 2,4D and Gramoxone. The manual cutting or removal of parasite from infected plants is also suggested.

5

Layout and Planting of Orchard

The plan reviewing the arrangement of fruit plants in an orchard is known as the orchard layout. Although several systems of planting are followed to increase the production of orchard, selection of root stocks, soil, system of training pruning and cultivars are very important factor influencing designing of layout. However, improper system results in overlapping of plant parts, competition for water, light, nutrient and unequal distribution of water etc., which ultimately results in poor performance or fruit orchard.

Layout plan

The marking of position of the plant in the field is referred as layout. The layout plan of the orchard should be prepared carefully, preferably in consultation with horticultural experts or fruit scientists. The orchard layout plan includes the system of planning provision for orchard paths, roads, water channels and farm building. A sketch of the proposed orchard should be prepared before the actual planting is taken up.

Aims of layout plan

1. To provide adequate space to plants.
2. To accommodate more number of plants.
3. Easy intercultural operations.
4. System of planting.
5. Selection of fruitcrops.

The following are the important systems of planting generally followed on the basis of agroclimatic conditions.

Material required for establishment of new orchard: Ropes, measuring tape, wooden pegs, wooden hammer, carpenter's square.

Types of layout

The plan showing the arrangement of plants in an orchard is known as the "orchard layout". There are several systems of planting and among them following are the important ones.

S.No.	Types of orchard layout	Calculation of plants per layout system
1	Square System	Area (sq. m) Row to row distance (m) x plant to plant distance (m)
2	Rectangular System	Area (sq. m) Row to row distance (m) x plant to plant distance (m)
3	Quincunx System	Number of plants in square system plus (+) one row less and one plant less in each row than the square system.
4	Hexagonal System	Number of plants in a square system plus (+) 15 per cent more than the square system.
5	Triangular System	Number of plants in square system minus (-) one plant less in every second row.
6	Contour System	

A. Square System

In this system row to row and plant to plant distances are kept similar. The plants are planted exactly at right angle at each corner. It is very easy method. Thus, every four plats make one square. Intercultural operations can be done in both directions as the distances between trees and rows are similar (5 x 5 m) e.g. Mango, Banana and citrus crops.

Advantages

1. Irrigation channels and paths can be made straight.
2. Operations like ploughing, harrowing, cultivation, spraying and harvesting becomes easy.
3. Better supervision of the orchard is possible as one gets a view of the orchard from one end to the other.

Disadvantages

1. Comparatively less number of trees are accommodated in the given area.
2. A lot of space in the centre of each square is wasted i.e, certain amount of space in the middle of four trees is wasted.

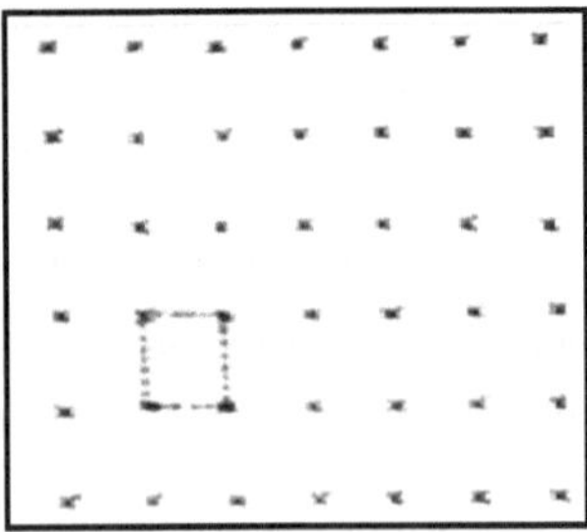

Rectangular system of planting

B. Rectangular system

This system is similar to that of the square in its layout except for the difference that the spacing between the rows and between the plants in a row is not equal. Thus, rectangular system accommodates more plants in rows. Inter-cultural operations can be carried out through both ways. The plants get proper space and sunlight for their growth and development. e.g., grapes (3 x 2 m).

Advantages

1. Intercultural operations can be carried out easily.
2. Irrigation channel can be made length and breadth wise.
3. Light can penetrate into the orchard through the large inter spaces between rows.
4. Better supervision is possible.
5. Intercropping is possible.

Disadvantages

1. A large area of the orchard between rows is wasted if intercropping is not practiced.
2. Less number of trees are planted.

C. Quincunx system

This system is also known as filler or diagonal system. This is a modification over Square system of layout. This system makes use of the empty space in the centre of each square by planting another plant. The plants that are planted in the centre of each square along with tall growing plants at the corners of squares are termed as "filler" plants. Generally, filler trees will be of short duration and not be of the same kind as those planted on the corners of the square. When main plants of the orchard resume their proper shape, the filler plants are uprooted. Guava, Peaches, Papaya etc. are important filler plants. In this layout population becomes double than square system of mango+ papaya, mango+ fig.

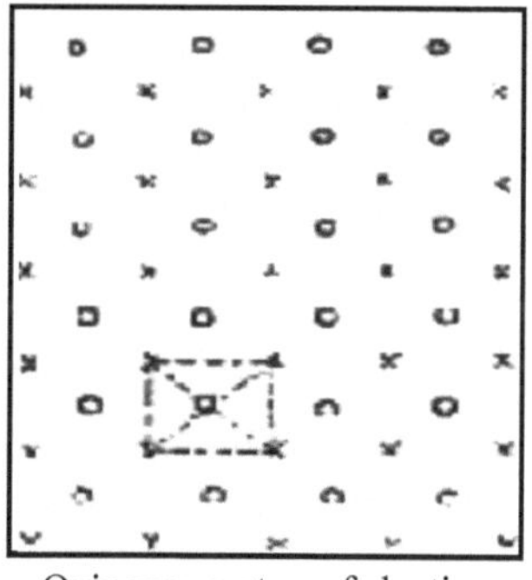

Quincunx system of planting

Advantages

1. Additional income can be earned from the filler crop till the main crop comes into bearing.
2. Compared to square and rectangular systems, almost double the number of trees can be planted initially.
3. Maximum utilization of the land is possible. Approximately 10% more plants than the square method.

Disadvantages

1. Skill is required to layout the orchard.
2. Inter/filler crop can interfere with the growth of the main crop.
3. Intercultural operations become difficult.
4. Spacing of the main crop is reduced if the filler crop is allowed to continue after the growth of the main crop.

D. Hexagonal / Triangular system

This system accommodates 15% more plants than square system. The plants are planted at the corner of equilateral triangle. For triangular planting, the markings are set as in the square or rectangular systems, except that those in the even numbered rows are mid-way between, instead of opposition in old pattern. Thus, six trees are planted making a hexagon. Seventh tree is planted in the centre. This is very intense method of planting and hence requires fertile land. In the suburb of cities where land is costly, this system is worth adoption. However, the laying out of system is hard and cumbersome.

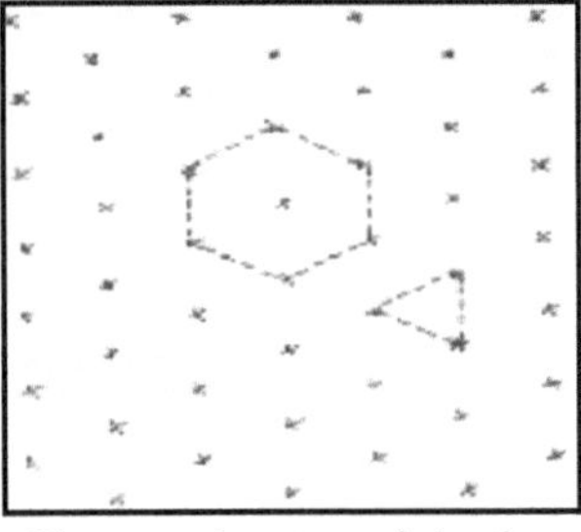

Hexagonal system of planting

Advantages

1. Compared to square system 15% more trees can be planted.
2. It is an ideal system for the fertile and well irrigated land.
3. Plant to plant distance can be maintained the same.
4. More income can be obtained.

Disadvantages

1. Intercultural operations become difficult.
2. Skill is required to layout the orchard.

Layout procedures

Establish a base line on one side of the field as in the square system.

Mark the position of trees on the base line at the desired distance and fix the Stakes. Make equilateral triangles on the base line maintaining the sides of the triangles equal to plant to plant distance.

Mark all the triangles with stakes and join them into a line to form the second line of trees. Similarly, make equilateral triangles on the second line and cover the whole land.

E. Contour system

It is adopted in hilly areas for planting fruit plants where land is undulated and soil erosion is a great threat. The layout is started from the lowest level and the tree rows are planted along a uniform slopes at right angle to the slope with a view to reduce loss of top-soil due to erosion. The width of contour terrace varies according to the slope of the hill.

Advantages

1. This system can be adopted in hilly regions, can control the soil erosion and helps simultaneously in the conservation of water.
2. Preservation of plant nutrients which are supplied as manures and fertilizers.

Disadvantages

1. Laying out of contour lines is difficult and time consuming.
2. Special skill is required to layout this system.
3. Special instruments are required for making contour lines.
4. The row to row distance will not be equal and adjustments may be required in the plant to plant distance.
5. Rows are broken in to bits and pieces.

Layout planning in minor fruit plants

The land must be cleared and all old trees and their roots removed - do not leave old timber lying around as this attracts pests. With land up to 15% slope, run the rows across the slope making sure there is a fall of 1 to 2% for drainage. When land is greater than 15% slope, contour planting must be undertaken. Suppose a plant is planted in rows 2 m apart with plants 1.5 m apart within the row. To mark the planting holes at this spacing on sloping land, follow the steps below. (1) Construct a simple wooden A-frame structure measuring 1.5 m high with legs 1.5 m apart. (2) The horizontal support cross-piece is marked at the central point. A string with a weight (stone or metal object) is attached at the apex of the 'A' and allowed to hang freely, similar to a pendulum.

Planting distance of major and minor fruit plants

The spacing given to the major and minor fruit trees is decided by the different factors *viz*. size of canopy, types of varieties, climatic condition, types of soil, growth habit of tree, selection of rootstock, facilites of irrigation and pruning technique etc. for better growth and development of fruit tree.

1. Best spacing regulates the proper utilization of sunlight, proper penetration of sunlight, avoids competition in the uptake of soil nutrients caused by the collision of root systems and facilitates proper irrigation requirements.
2. The latest technology on high-density plantation system where trees are planted at small unit areas for maximum utilization of space is becoming popular and more number of plants were planted
3. It will be very difficult to suggest exact spacing for fruit trees, which will suit every locality or soil.

Given below is the spacing of some of the important fruit plants, which serve as basic guideline for establishing a new orchard.

Planting distance of major and minor fruit plants

S.No.	Name of fruit tree	Planting distance (m)	Number of plants /ha (square system)
1.	Mahua	10x10	100
2.	Ber	6.5 x6.5	236
3.	Aonla	8x8	156
4.	Phalsa	2.5 x2.5	1600
5.	Karonda	2.5 x2.5	1600
6.	Fig	5x5	400
7.	Mango	10	100
8.	Citrus & Pomegranate	6	275
9.	Grape i) Head system ii) Kniffin system iii) Bower system	2.0x1.54.00x 3.00 3.1x6.0	3300, 1100 and 550
10.	Guava, Peach, plum& Loquat	6.5x6.5	225
11.	Litchi	7.5	177
12.	Pear	7.5	180
13.	Date palm	6	277
14.	Jamun	7.5	180
15.	Sapota	6x6	277
16.	Almond	7.5x7.5	177
17.	Apple	4x4	625
18.	Mangosteen	7 x 7	204
19.	Papaya	2x2	2500
20.	Mulberry Black mulberry	7 x 7	204
21.	PearPunjab Nectar (Low chilling)	5 x 5	400
22.	Mandarin Kinnow	5 x 5	400
23.	Kiwi fruit Kiwi fruit Female: Abbott, Bruno, Hayward, MontyMale: Tomuri, Allison, Matua	6 x 6	277

6

Plant Propagation and Its Methods

Plant Propagation

Plant propagation involves the formation and development of new individuals, which are used in establishment of new plantings. It is simply the reproduction or multiplication of a plant from a source that is often referred to as a mother plant. In general, two methods are employed: 1) sexual and 2) asexual. Sexual propagation is multiplication of plants from seed and asexual or vegetative propagation involves starting a new plant from some vegetative part of a plant. Propagation by seed results in seedling variability. It is an extreme disadvantage for a nursery operator trying to produce a uniform crop containing chosen desirable characteristics. Seed propagation is not practiced in fruit nursery as plants raised by this method fail to bear true to type fruits.

Germination of seed

The activation of the metabolic machinery of the embryo leading to the emergence of a new seeding plant is known as germination. Germination is essentially a quickening of the growth of the embryo. As germination proceeds, the growing points of the radical and plumule divide rather rapidly. Usually the radical emerges from the seed coat first, proceeds downward, and develops into the root system, the plumule proceeds upward and develops into the shoot system. Germination is entirely a food utilization process. Processes going on in seed during germination are: 1) Absorption of water; 2) Secretion of enzymes and hormones; 3) Hydrolysis of stored food into soluble form; and 4) Translocation of soluble foods and hormones to the growing points. These processes are either wholly or in part influenced by the food reserves, hormone supply, water supply, oxygen supply and temperature level.

Advantages of sexual propagation

1. This is very simple and easy method of propagation.
2. Some species of fruit crops which cannot be propagated by asexual means should be propagated by this method. e.g. Papaya, Karnoda, Khejri, Ker and Phalsa.

3. Hybrid seeds can be developed by this method.
4. New variety of crops is developed only by sexual method of propagation.
5. Root stocks for budding and grafting can be raised by this method.
6. The plants propagated by this method are long lived and are resistant to water stress.
7. Transmission of viruses can be prevented by sexual method.
8. Seed can be transported and stored for longer time for propagation.

Disadvantages of sexual propagation

1. Characteristics of seedling propagated by this method are not genetically true to type to that of their mother plant and it requires long period for fruiting
2. Plants grow very high, so they are difficult for intercultural practices like spraying,weeding, training and pruning and harvesting etc.
3. The plants which have no seeds cannot be propagated by this method. e.g. Banana, pomegranate strawberry, kiwi and fig.

Vegetative propagation

Vegetative propagation of fruit trees dates back to ancient times. The Greeks and Romans adopted this strategy and spread these methods all over Europe.Plant can be asexually reproduced either by using part of two or more plants in a union or parts of the same plant.Asexual propagation may be done by making cuttings from the stem, root or leaves of the desired plant. Stem cuttings are made by removing a small branch or twig from the plant. This cutting will usually contain two or more buds, one of which will grow into the top of the plant. With proper treatment, Adventitious Roots will be produced on the end of the cutting that was closest to the root of the original plant. Root cuttings are made in a similar fashion, but produce an adventitious stem on the end of the cutting that was nearest to the stem of the original plant. Leaf cuttings produce both roots and stems when the leaf is placed under proper conditions. Asexual propagation is used to reproduce or multiply many horticultural plants. Plants that are propagated asexually are genetically the same as the mother plant. This is also called cloning. Some asexually propagated crops that are grown extensively are: tree fruits, strawberries, banana, cranberries, and most herbaceous and woody ornamental plants.

Asexual propagation is called with different names — Asexual propagation, Vegetative propagation, Clonal propagation. Asexual propagation is reproduction by means of vegetative parts of the plant such as roots, shoots, or leaves other than seed. In this propagation sexes are not involved–hence it is called asexual propagation.

i. Propagation by apomictic seedling e.g., mango, apple and citrus.

ii. Propagation by specialized vegetative structures

   1. Propagation by division: Division of rhizome, e.g., Banana and blueberry etc.

   2. Propagation by suckers: Pineapple, banana, blackberry and raspberry etc.

   3. Propagation by runners: Strawberry

   4. Propagation by offset: Pine apple and banana etc.

Scion: That part of the union to be attached to the rootstock.

Stock: That part of the union which contains the root portion of the union

Inter-stock: Inter-stock is a piece of stem inserted by means of two graft unions between the scion and rootstock. Inter-stocks are used to avoid an incompatibility between the rootstock and scion, to produce special tree forms, to control disease or to take advantage of its growth controlling properties.

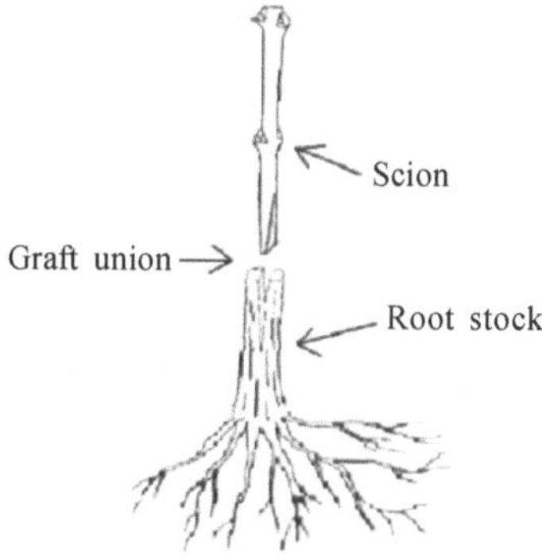

Vascular cambium: Vascular cambium is a thin tissue located between the bark and the wood. Its cells are meristematic that is they are capable of dividing and forming new cells. For successful graft-union, the cambium of the scion is placed in the close contact with the cambium of the rootstock.

Callus: Callus is a term applied to the mass of parenchyma cells that develop from and around wounded plant tissues. It occurs at the junction of a graft union, arising from the living cells of both the scion and rootstock. The production and interlocking of this parenchyma (callus) cells constitute one of the important steps in callus bridge formation between the scion and rootstock in a successful graft.

Topworking: Applies to the process of changing the top of a plant from one cultivar to another by grafting or budding. This procedure may sometimes involve a series of multiple grafts.

Cambium: It is a layer of dividing cells in a stem that is responsible for increasing the stem diameter. Plants lacking cambium (example: monocots such as corn) cannot be grafted. The cambium of a stock and scion must be in close contact to form a union. Cambial activity during spring facilitates easy separation of bark from the wood.

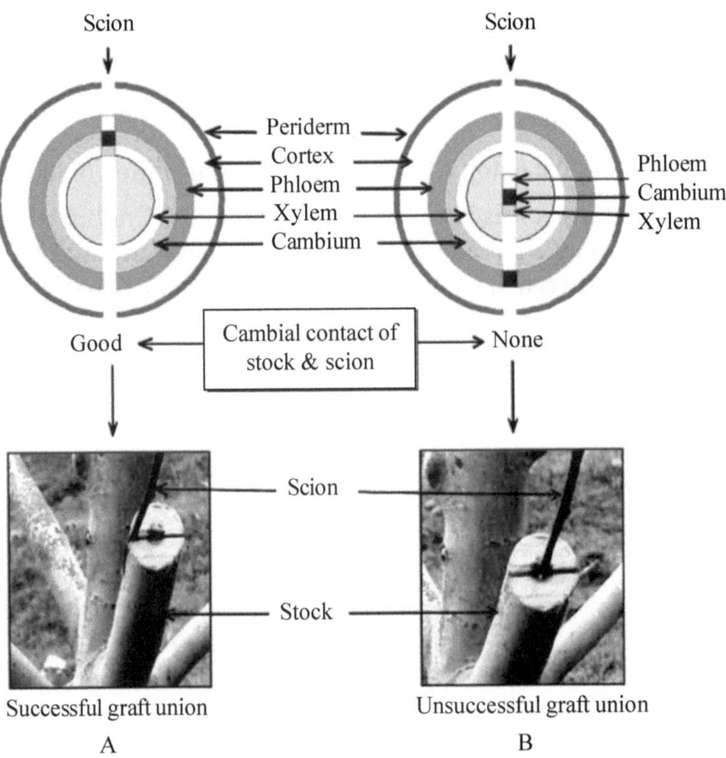

Sap: It is the fluid transported via conductive tissues such as **xylem** and **phloem**. While **xylem** transports water from roots to the aerial parts of the plant, **phloem** conducts sugars, nutrients and hormones from the leaves to the roots and storage organs (fruits). During the selection of the scion wood some important points to keep in mind.

The scion wood must carry healthy buds that will grow into leafy shoot-It should come from a tree which is free from any pests or diseases. The plant should have the required characteristics. This includes, that we take scions from plants, which is already bearing fruits.

Grafting tools

The essential tools are pruning shears, budding/ grafting knives, saws, sharpening stone etc.

1. Grafting

Joining the parts of two separate plants so that they will unite and continue to grow as a single plant.

S. No.	Types of apical grafting	Side grafting	Bark Grafting	Root Grafting	Approach Grafting	Repair grafting
1	Whip-and–tongue grafting	Side–veneer grafting	Bark graft (rind grafting)	Whole-root and piece root grafting	Spliced approach grafting	Inarching
2	Splice graft (whip grafting)	Side-tongue grafting	Bark graft (rind grafting)		Tongued approach grafting	Bridge grafting
3	Cleft-graft (split grafting)	-	-	-	Inlay approach grafting	Bracing
4	Wedge graft (saw-kerf grafting)	-	-	-	-	-
5	Saddle grafting	-	-	-	-	-
6	Four flap graft (banana grafting)	-	-	-	-	-
7	Whip-and – tongue grafting	-	-	-	-	-

Methods of grafting

a. **Whip grafting :** Whip grafting method is useful for grafting relatively thin materials i.e 0.8-1.2 cm diameter and is highly successful due to considerable cambial growth.

b. **Cleft grafting**: Cleft grafting is used for top working or more than one year rootstock and should be completed before active growth of the stock. Scions are usually about 0.25-0.30 inch in diameter and have 2 to 3 buds. Stocks should be 2 to 4 inches in diameter and straight. Make the cut so there are 4 to 6 inches below with no knots or side branches. Drive the tool in with a wooden mallet. Remove the cutting edge of the clefting tool and drive the wedge part of the tool in the center of the stock to open the split to receive the scions. Wedge open the branch after the scions have been prepared to avoid drying of the freshly split tissue. This method is commercially used in walnut, mango and almond crops.

c. **Side-veneer graft**: Make a shallow cut, about 1.5 inches long at the base of the stock, directed slightly inward. At the base of this cut, make a short inward, downward cut to intersect the first cut, thus allowing removal of a piece of wood and bark. Prepare the scion with a long cut the same length and width as that of the first cut on the stock. Make a short cut on the opposite side of the base of the scion. Insert the scion in the stock with the long cut of the stock next to the long cut on the scion. Secure the scion by wrapping with tape or rubber budding strips. Cover the graft region with a protective material.

d. **Inarching grafting**:Inarching is a repair technique that may be used instead of bridge grafting. Suckers growing at convenient locations next to the injured area, seedlings, or rooted cuttings planted next to the trunk are bark grafted into the trunk above the injured area. Cover the graft area with a protective material. Remove shoots which develop on the inarches. If the tree is on a dwarfing rootstock, the rootstock used for inarching should be of the same kind (otherwise the benefits of the original rootstock will be lost).

e. **Approach grafting**: Approach grafting means adjacent of two branches are joined together. For two plants on their own roots, the main stems are joined together. This method has the advantage of an uninterrupted flow of water to the scion from its own roots until the union is formed. Likewise, the rootstock receives manufactured food from its top during the graft union formation. Stock and scion or adjacent branches within a tree may be joined by the spliced method which uses single, long, smooth cuts on adjacent surfaces.

2. Propagation by cutting

a. **Stem cutting :** Stem cuttings are the most convenient and popular method of plant propagation. A stem cutting is any cutting taken from the main shoot of a plant or any side shoot growing from the same plant or stem. It is essential for the cuttings to have a sufficient reserve food to keep tissue alive until root and shoot are produced. The shoots with high carbohydrates content roots better. Cuttings from young plants root better, but, if older shoot of the plant are cut back hard, very often they can be induced to produce suitable shoots for rooting. There are several types of stem cuttings.

b. **Hardwood cutting :** Hard wood cuttings are prepared during dormant season, usually from one year old immature shoots of previous season's growth. The length of cuttings varies from 10-15 cm in length and 0.5 to 1.5 cm in diameter, depending upon species. Each cutting should have at least two buds. While preparing the cutting, a straight cut is given at the base of shoot below the node while a slanting cut 1-2 cm above the bud is given at the top. e.g. grape, fig, pomegranate, mulberry and kiwifruit etc. Spring and autumn season are the best time for planting the cuttings. Different growth regulators may be used to promote rooting. Good response in cuttings can be obtained by overnight dipping in 100-300 ppm IBA. 2000 to 5000 ppm of indolebutyric acid (IBA), indoleacetic acid (IAA) and nephtalenacetic acid (NAA) through quick dip methods are the most successful growth regulators.

c. **Semi hardwood cutting :** Semi hardwood cuttings are prepared from partial matured, slightly woody shoot. These are succulent and tender in nature and are usually prepared from growing wood of current seasons growth. The length of cutting varies from 7-20 cm. The cuttings are prepared by trimming the cutting with straight cut below a node and removing a few lower leaves. However, it is better to retain two to four leaves on the top of cuttings. Treating the cutting with 2000 to 4000 ppm IBA before planting gives better results e.g., guava, jackfruit and lemon.

d. **Softwood cutting :** Softwood cutting is the name given to any cuttings prepared from soft, succulent and non-lignified shoots which have not become hard or woody. Usually the cutting size is 5-7.5 cm but it varies from species to species. Usually some leaves are retained and before planting, treatment with auxin (IBA) is beneficial. e.g. lime and lemon.

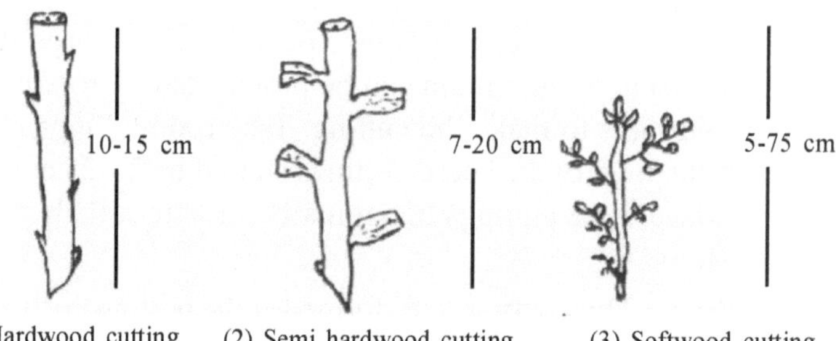

(1) Hardwood cutting (2) Semi hardwood cutting (3) Softwood cutting

e. **Leaf bud cutting :** Leaf cutting should preferably be prepared during growing season because buds if enter in dormancy may be difficult to force to active stage. A leaf bud cutting consists of a leaf blade, petiole and shoot piece of stem with attached axillary bud of active growing leaves. In this cutting, 1-1.5 cm stem portion is used when propagating material is small eg. Blackberry, lemon, raspberry.

Propagation system with different types of cuttings

Cutting type	Hardwood (deciduous)	Hardwood (evergreen)	Semi-hardwood	Softwood	Leaf	Root
Description	Mature, dormant, quiescent	Mature hardwood stems, woody spp.	Partially mature on current season	New, soft succulent	Leaf blade and petiole	Root piece from thin flesh roots
Season propagated	Dormant season: late fall to early spring	Dormant season: late fall to late winter	Late spring to late summer	Spring to early summer	Year round	Late winter or early spring
Propagation system	Field and green house propagated	Light mist, fog, humidity,	Light mist, fog, humidity,	Light mist, fog, humidity,	Light mist, fog, humidity,	Depending on spp.
Cutting length	10-26 cm, at least 2 nodes with basal cuts just below the node	10-20 cm (4-8 inches)	7.5-15 cm (3-6 inches)	7.5-12.5 cm (3-5 inches)	Varies with spp. and leaf size.	Small, delicate roots-(2.4-5 cm), large roots (5-15 cm)
Chemical treatment	IBA or NAA at 2,500-5,000 ppm.10,000 ppm max. with difficult to root	IBA or NAA at 2000 ppm or slightly high. Difficult to root; 5000-10,000 ppm	IBA or NAA at 1,000-3000 ppm. max. 5000.	IBA or NAA at 500-1,2500 ppm ,max 3,000ppm	Cytokinin is applied. 100-230 ppm	Usuallyno hormone is applied
Example of plants	Apple, pear, plum, quince rootstock	-	olive	Peach, pear, plum	blackberry	

f. **Root cutting:** Apple, cherry, pear, pecan, guava, breadfruit, apple, blackberry, raspberry and plum can be propagated by root cutting. Select 2-3 years old roots to make the cutting. Take root cuttings about 1 meter away from the tree trunk. These cuttings should be 20-25 cm long and 1-2 cm thick. Place these cuttings horizontally into the soil about 10 cm deep until they shoot.

g. **Torn Cuttings:** This cutting is performed at the bottom portion of the stem where there is a union with the mother plant. This is a very old technique and it is rarely used now-a-days.

h. **Hammer Cuttings:** In this technique, a piece of twig is cut together with the stem. Some plant cuttings, like gooseberry cuttings, are difficult to root and the additional piece of twig helps to develop root system.

i. **Semi-Wooded Cuttings:** These types of cuttings are usually made from woody evergreen plants, which are taken during the growing season. They are cut off before the wood hardens and turns brown. Cuttings are used from the leafy shoot tip. When the cuttings have developed their root systems, we can then transplant each one into a larger container. The propagation technique is used for the reproduction of coffee, kiwi, litchi, macadamia, mango, granadilla and pomegranate plants.

j. **Truncheons:** Truncheons are branches, about as thick as a human arm that we can grow into new plants. The branches are about 170-180 cm long. Cut the top of the branch at a slant, which prevents water from rotting the trunk cheon. Truncheon should be planted into a narrow hole about 60 cm deep. The best time for this method is the end of the dormant season when the plant still grows slowly.

3. Layering methods

The main purpose of layering is to provide rooting for the stem of the mother plant. The new growing plant will keep the union with the mother plant until it is able to survive on its own. Different techniques exist for plant propagation through layering methods.

Type of layering

a. Mound layering

This is the most common method which is used to propagate pear, quince and apple rootstock. This types of layering involves heading back an established plant close to ground level during the dormant season.

b. Simple Layering

Simple layering technique is commonly used for hazel-nut propagation. During the dormant season, stems are bent down into a 20-25 cm deep trench and covered with soil. The top parts of the stems, which usually have 2-3 buds on them, remain above the surface.

c. Air Layering

Air-layering or 'gootee' is widely accepted method of litchi, guava, jackfruit and kagzi lime propagation. This is the most widely used propagation method and the one, which gives the most satisfactory results during rainy season. Its major advantages are that it is simple to use and genetically identical plants are produced. The most serious drawback of air layering is the damage to the parent plant if a large number of layers are required and poor survival in the nursery after shifting of layers. However, plants prepared by this method are delicate and difficult to transport. Air-layering is done when leaves of the previous growth flush have proper maturity. Even though, it has been reported that the better the branch used, the better the root system obtained. Excellent results can be obtained with branches of 10-25 mm diameter and 46-60 cm length. In this method a ring of bark, about 2 cm wide just below a bud, is removed from a healthy and vigour Twig of about one year old and 2.4 - 4 cm in diameter. The cut is then surrounded by Sphagnum moss grass wrapped with polythene sheet. These shoots produce rooting rate above 90 per cent and damage to the mother plant is also minimal. Air layering can be done at any time of year as long as there is sufficient moisture, but best results are obtained in rainy and spring season. Branches should be selected on the periphery of the trees, so that they can easily be worked on.The selected branch should consist of a single stem and other stems must be removed.

4. Budding methods

Budding is a modification of grafting, involving the use of a scion with only a single bud attached to a piece of bark. The method of budding is the most common technique for plant propagation in commercial nurseries. The fruit plants like apple, pear, peach, ber, aonla, bael, jamun, plum and other which are easily multiplied through budding methods with good successes. First, one must graft a single bud attached to the stem of the rootstock. However the budding operation in deciduous fruit plants starts from end April to May beginning when the spring growth flush gets somewhat hardend and is suitable for taking bud stick and subtropical plants starts during rainy season. The stem or branch may not be thicker than 2 cm diameter. Therefore, this method is only applicable for young rootstock plants or smaller branches of large plants. For best results, use bud wood or bud sticks which are of a vigorous current season growth. Remove the top and bottom part of the branch, because the tip buds are too immature and the bottom buds may be a cluster of buds or they are too weak to use for budding. The length of the stick is approximately 30 cm. Remove the leaves leaving a 1-1.5 cm long of leaf petiole on the stem. The time for budding comes when the bark peels easily on the stock. Irrigation a few days before budding helps to slip the bark.

Budding techniques

The different methods of patch budding, forket budding used for temperate fruit crops, whereas T or shied, inverted T, Ring budding etc., which are used under multiplication of evergreen fruit plants.

Selection of bud-wood

One should use vegetative buds than the flowering buds for budding. The vegetative buds are usually small and pointed while flower buds are large and plump. In case of bud-wood to be procured from distant place the leaves must be removed by leaving petiole intact. The bundle of moist bud wood should be packed in moist jute/gunny bags, cloth and moistened in transit after 6-8 hours if possible or wrapped with polythene. In general, the success of a particular budding method depends upon the bark's slipping ability of stock and scion. Among the different methods, chip budding can only be done when bark is not slipping.

Advantages of budding

- It is the best propagation method if the propagating material is scarce.
- Budding is useful in plants, which release excessive wound gum (e.g. stone fruits) from injury carried to wood portion of the stem at the time of grafting.
- Budding union is stronger than grafting so damage by wind or storm is less.
- Simple, efficient and quicker method of propagation.

Type of budding

a. T budding

The "T" cut on the stock is done about 20-25 cm above the surface with a 2 cm long vertical cut and a 7-8 mm long horizontal cut on the stock. A slight twist with the budding knife may open the two flaps of bark. After that, the bud should be inserted under the two flaps of bark by pushing downward. If part of the bud remains above the horizontal cut, it must be cut off. This will allow the flaps to be closed tightly. Finally, the incision should be closed with budding tape, which should be wrapped tightly around the stem. Tying must start at the bottom or the top end of the incision. After 3-4 weeks, the tape should be removed. At this time, the shield of the bud and the petiole may indicate the condition of the bud. With high labour costs, only a few of the more efficient grafting techniques are utilized, such as side veneer, tongue and whip-tongue grafts. Chip budding and T-budding are two most common budding methods used in fruit crops and ornamental plants. In automated and manual grafting of herbaceous vegetable crops, the splice graft is one of the most important.

b. Patch budding

In case of patch budding, a rectangular patch of bark is removed completely from the rootstock and replaced with a patch of bark of the same size containing a bud of the cultivar to be propagated. It is slower and difficult to perform than T-budding. It is widely used in thick-barked species, such as walnuts, pecans and rubber tree, where T-budding gives poor results due to poor fit around the margins of the bud-particularly the top and bottom. It is usually done in late summer or early fall, but can be done in spring also. In patch budding, the stock and scion should preferably of same thickness (20-25 mm). First a rectangular piece of bark (25mm long and 10-15 cm wide) is removed from the stock and a similar patch, containing a bud is removed from the scion by making two horizontal cuts above and below the bud and then two vertical cuts connecting the horizontal cut. After removing the patch, the bud should fit tightly at the top and bottom. It is then wrapped with polythene strip, keeping the bud uncovered. The wrapping material should hold the bark tightly and cover all the cut surfaces to prevent free entry of air or water. After the bud starts sprouting, the stock above the bud union may be cut off step by step. In addition to pecanut and walnut, mango, rubber plant, aonla, jackfruit and jamun are also propagated by this method.

c. Ring or annular budding

In this type of budding, a complete ring of bark is removed from the stock and it is completely girdled. A similar ring of bark containing a bud is removed from the bud stick and is inserted on to the rootstock. The thickness of stock and scion should be same size. It has been utilized in ber, peach and mulberry because the newly emerged shoots from the heavily pruned plants are capable of giving such buds for budding, which can be easily separated. In this method since the stock is completely girdled and if the bud fails to heal in, the stock above the ring may eventually die.

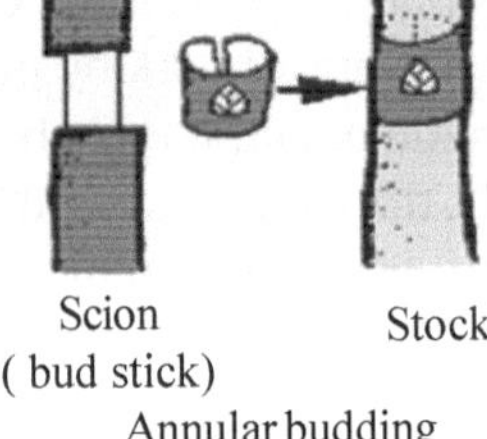

Annular budding

d. Forket budding

In forket budding, the stock is prepared by giving two vertical cuts and a transverse cut above the vertical cuts to join them. The bark is removed carefully along the cuts, so the flap of bark hangs down. The scion is prepared in a fashion similar to patch budding, having the size similar to cuts made on the stock. The scion is then slipped into the exposed portion of the stock and the flap is drawn over the inserted bud patch. It is then tied with a suitable wrapping material. After successful growth of bud, the portion of stock above union is removed carefully.

e. Flute Budding

In flute budding, the bark patch from the stock is removed in such a way that it almost completely encircles the stock except with a narrow bark connection between the upper and lower cuts on the stock. A similar patch of bark is removed from bud stick containing a healthy bud. The shield containing the bud is then inserted in the vacant area of the stock; the shield should fit tightly on the stock. It is then wrapped with suitable wrapping material, leaving the bud uncovered. The other procedure is same as in patch budding. Because of presence of a narrow connecting strip of bark on the stock, it remains alive even if the bud fails to sprout.

f. Micro-budding

Micro-budding is used successfully for propagating citrus particularly in Australia. It is similar to "T" budding except that the shield (bud piece) utilized is thin and tiny like "T-budding"; the micro budding is also not done under aseptic conditions. The petiole is cut off just above the bud and then bud is removed from the bud stick by a flap cut just underneath the bud. Thus, only the buds are utilized in micro-budding. In stock an inverted "T" cut is made and the tiny shield containing the bud is inserted in it and later tied with a thin plastic tape. The tape may be removed soon (15-20 days) after the healing has taken place.

Propagation system with different types of cuttings

Cutting type	Hardwood (deciduous)	Hardwood (evergreen)	Semi-hardwood	Softwood	Leaf	Root
Description	Mature, dormant, quiescent	Mature hardwood stems, woody spp.	Partially mature on current season	New, soft succulent	Leaf blade and petiole	Root piece from thin flesh roots
Season propagated	Dormant season: late fall to early spring	Dormant season: late fall to late winter	Late spring to late summer	Spring to early summer	Year round	Late winter or early spring
Propagation system	Field and green house propagated	Light mist, fog, humidity,	Light mist, fog, humidity,	Light mist, fog, humidity,	Light mist, fog, humidity,	Dependin g on spp.
Cutting length	10-76cm, at least 2 nodes with basal cuts just below the node	10-20cm (4-8 inches)	7.5-15cm (3-6 inches)	7.5-12.5 cm (3-5 inches)	Varies with spp. and leaf size.	Small, delicate roots-(2.4-5cm), large roots (5-15cm)
Chemical treatment	IBA or NAA at 2,500-5,000 ppm.10, 000 ppm max. with difficult to root	IBA or NAA at 2000ppm or slightly high. Difficult to root ;5000-10,000 ppm	IBA or NAA at 1,000-3000 ppm.max.50 00.	IBA or NAA at 500 - 1,2500 ppm ,max 3,000ppm	Cytokinin is applied. 100-230 ppm	Usually no hormone is applied
Example of plants	Apple, pear, plum, quince rootstock	-	olive	Peach, pear, plum	blackberry	

Recommended propagated method of fruit crops

S.No.	Fruit Crops	Commercial Propagation Techniques	Type of Rootstock
1.	Almond	T-budding and wedge grafting	Open pollinated peach and bitter almond seedlings
2.	Aonla	Patch budding and wedge grafting	Open pollinated seedlings
3.	Apple	T-budding/tongue and wedge grafting	Clonal rootstocks
4.	Apricot	T-budding and wedge grafting	Open pollinated apricot/peach/ plum seedlings
5.	Avocado	T-budding and wedge grafting	Open pollinated seedlings
6.	Bael	Wedge grafting	Open pollinated seedlings
7.	Banana	Suckers/corm	An elite mother plant, free from diseases,
8.	Cashew	Soft wood grafting	Open pollinated seedlings
9.	Cherry	Tongue and wedge grafting	Clonal rootstocks
10.	Custard apple	Wedge grafting	Open pollinated seedlings
11.	Date palm	Sucker/off shoot	An elite mother plant, free from diseases, producing
12.	Fig	Hard wood/semi-hard wood cutting	An elite mother plant, free from diseases
13.	Grape	Hard wood cutting and wedge	
14.	Guava	Wedge grafting	Open pollinated seedlings
15.	Gooseberry	Hardwood/semi-hard wood cutting	An elite mother plant, free from disease
16.	Jackfruit	Patch budding and soft wood grafting	Open pollinated seedlings
17.	Jamun	Soft wood grafting	Open pollinated seedlings
18.	Khirnee	Soft wood grafting	Open pollinated seedlings
19.	Kiwi fruit	Hard/semi hard wood cutting and	
20.	Lime	Cutting/ layering	
21.	Litchi	Air layering and wedge grafting	Open pollinated seedlings (for wedge grafting)
22.	Mahua	Soft wood grafting	Open pollinated seedlings
23.	Mandarin	T-budding and wedge grafting	Polyembryonic seedlings
24.	Mango	Soft wood, wedge and veneer grafting	
25.	Peach	T-budding, wedge and tongue grafting	Clonal rootstocks
26.	Pear	T-budding, wedge and tongue grafting	Clonal rootstocks
27.	Pecan nut	Patch budding and wedge grafting	Open pollinated seedlings
28.	Pineapple	Slip/sucker	An elite mother plant,
29.	Plum	T-budding, tongue and wedge grafting	Clonal rootstocks
30.	Pomegranate	Wedge grafting/air layering	Open pollinated seedlings
31.	Sapota	Wedge grafting	Khirnee seedlings
32.	Blackberry / Resberry	Sucker	-
33.	Strawberry	Runner	
34.	Sweet Orange	T-budding/wedge grafting	Polyembryonic seedlings
35.	Wood apple	Soft wood grafting	Open pollinated seedlings
36.	Walnut	Patch budding and wedge grafting	Wild seedlings and Paradox
37.	Sweet Orange	T-budding/wedge grafting	Polyembryonic seedlings

Table 1: Propagation methods commonly used for selected tropical fruit crops (Oliver 1903, Garner and Chandri 1976, Wasielewski and Balerdi 2016, and other sources).

S. No	Fruit Suckers	Scientific Name	Seeds	Cuttings	Air	Grafting Layer	Budding Runners
1.	Abiu	*Pouteria caimito*	X		XX		
2.	Apple		X		XXX		XX
3.	Apricot		X		XXX		XX
4.	Avocado	*Persea* spp.	X		XXX		
5.	Banana	*Musa* spp.			XXX		
6.	Beal		X	X	XX		
7.	Bilimbi	*Averrhoa bilimbi*	X		XX		
8.	Breadfruit	*Artocarpus altilis*	~~X~~	X, Roots	X		
9.	Cacao	*Theobroma cacao*	X	Stem,X	X	XX	
10.	Carambola/ Starfruit	*Averrhoa carambola*	X	X	X	XX	
11.	Cashew	*Anacardium occidentale*	X		XX		
12.	Cherimoya	*Annona cherimola*	X	X, Stem	X	XX	
13.	Chico/ Sapodilla	*Manilkara zapota*	XX	X	X		
14.	Citrus	*Citrus* spp.	X	X	X	XX	XXX
15.	Durian	*Durio zibethinus*	X	X, Stem	X	XX	
16.	Eggfruit/ Canistel	*Pouteria lucuma*	X		X	XX	
17.	Fig	*Ficus carica*	X	XXX	X	X	
18.	Green Sapote	*Pouteria viridis*	X	X		XX	
19.	Grumichama/ Brazil Cherry	*Eugenia brasiliensis*	XX	X		X	
20.	Guava	*Psidium guajava*	X	Cutting Root	X X	XX	X
21.	Jaboticaba	*Myrciaria cauliflora*	XX		X	X	
22.	Jackfruit	*Artocarpus heterophyllus*	X	X, Stem	X	XX	
23.	Jamun						
24.	Karnoda						
25.	Kinnow						
26.	Langsat/ Lansone	*Lansium parasiticum*	X	X	XX	X	
27.	Lilikoi/ Passionfruit	*Passiflora edulis*	X	X	XX	X	
28.	Litchi	*Litchi chinensis*	X		X X X	X	
29.	Longan	*Dimocarpus longan*	X	X	X X		
30.	Loquat	*Eriobotrya japonica*	X	X	X	XX	
31.	Mamey Sapote	*Pouteria sapota*	X		X	XX	
32.	Mango	*Mangifera indica*	X		X	XXX	
33.	Mangosteen	*Garcinia mangostana*	X	X, Stem			
34.	Marang	*Artocarpus odoratissimus*	XX	XX	X	X	
35.	Mulberry	*Morus* spp.		XX			

36.	Papaya	*Carica papaya*	XX X				
37.	Pear		X			XX	X
38.	Phalsa		XXX	X			
39.	Plum						
40.	Pomegranate	*Punica granatum*		XXX			
41.	Pulasan	*Nephelium mutabile*	X			XX	
42.	Rambutan	*Nephelium lappaceum*	X		XX	XXX	
43.	Rollinia	*Rollinia deliciosa*	X	X	XX		
44.	Soursop	*Annona muricata*	XX	X			
45.	Star Apple/ Caimito	*Chrysophyllum cainito*	X	X	X	XX	
46.	Strawberry XXXrunner			Y	Y	Y	
47.	Surinam Cherry	*Eugenia uniflora*	X		X	X	
48.	Tamarind	*Tamarindus indica*	XX	X	X	X	
49.	Water/ Wax Apple	*Syzygium samarangense*	X	XX	X	X	
50.	Wax Jambu	*Syzygium jambolana*	X	X	X	XX	

Notes

X = Some difficulty and successes rate less

XX = Possible with most common method

XXX = Commercially and common method

Y = Not possible

7

Orchard Management under Rainfed Condition

Selection of orchard soil in rainfed areas

A good or adequate soil for orchard production is to say that a given plot or area of land must possess the following qualities.

i. A deep and well aerated soil with good drainage.

ii. Adequate nutrients for optimum tree growth and high yields.

iii. Good Water holding capacity - the ability to store and feed the fruit tree with an abundant supply of water.

It is a scientific technique which provides maximum outputs of various inputs consistently without any loss of fertilizers and manure, plant, plant protection chemicals, produce etc. Therefore, one should understand the management of these qualities of both resource and output.

Important factors of orchard management

1. Soil Moisture and fertility management of waste/degraded land under rainfed areas
2. Water shed management under rainfed condition
3. Organic and inorganic fertilizer management
4. Plant canopy and weed management
5. Management of biotic and abiotic stress under rainfed condition.
7. Bearing, rejuvenation of old and senile orchards fruitfulness management.
8. Pre-post fruit dropping, maturity and harvest.
9. Pre cooling, post harvest handling, utilization and marketing.

1. Soil moisture and fertility management of waste/degraded land

a. To create favorable conditions for moisture supply through organic mulches, proper drainage and enhancing the catchment area under rainfed condition.

b. To maintain high fertility level though sod culture, FYM and replenishment against losses.

c. To provide proper soil conditions for gaseous exchange and microbial activities through addition of organic matter.

d. Soil erosion management.

e. To make sure supply of nutrients for growth and development of plants.

f. Utilization of inter row spacing and others vacant land for additional income because such a loss is inconceivable for small holders.

g. To minimize the cost of cultivation with high economic returns.

h. To suppress weed population.

Soil management of fruit orchard in rainfed areas

Cover crop: The crop grown to provide a cover to soil to protect it from erosion. It may be green manure crop also i.e. cow pea, urd and moong etc.

Intercrop: Any crop other than main crop grown between the rows of perennial trees is known as intercropand this type of cultivation is called as intercropping.Okra, Turmeric, pea, cucurbits, Kharif and rabi pulses and Phalsa fruit etc. are suitable for intercropping under rainfed orchards.

Green manure crop: The crop other than main crop grown for the purpose of enriching the soil for organic matter is called green manure cropi.e. Sesbina, kharif and rabi pulses etc.

Methods of soil management: Appropriate soil management method is important for the control of weeds, incorporation of organic and inorganic fertilizers and to facilitate absorption of water in soil.

i) The common soil management practices are

1) Cultivation, 2) Sod Culture, 3) organic and inorganic Mulches, 4) intercropping, 5) Rotation, 6) Rooting depth of the crop, 7) Slope of the soil, 8) increase water catchment area. Cultivation in context with soil management refers to working of the soil by ploughing, harrowing, disking or hoeing. It is essential for removal of weeds, incorporation of manures and fertilizers, green manuring and to facilitate water and nutrient absorption through better aeration. Depth of tillage and areas

are determined by root depth and spread of the canopy of the tree. In cultivation different modifications are made under specific conditions.

Cultivation of cover crops

- In areas where soil is eroded during rains and drainage is poor, soil is cultivated and cover crops are grown between the rows during rains.
- The crop may and may not be turned into soil.
- These crops not only increase water retaining capacity of soil and biological complex of the soil but also add organic matter when ploughed in besides checking erosion.
- As cover crops, legumes should be preferred because they add extra N in soil through fixation of atmospheric-N in their nodules.
- They also suppress weeds during rainy season.
- Crops like greengram, blackgram, cowpea, cluster bean, and soybean should be preferred during kharif season
- While pea, fenugreek, broad bean and lentil can be preferred in winter season as cover crops.

Advantages

- Adds organic matter in soil.
- Improves soil condition.
- Improves soil fertility.
- Increases water retention capacity of soil.
- Increases biological complexes of soil.
- Checks soil erosion.
- Checks nutrient losses through soil erosion.

ii) Cultivation and intercropping

Intercropping is growing of two or more crops simultaneously on the same field so that crop intensification occurs in both time and space dimensions and there is intercrop competition during all or part of crop growth. This can be mixed strip or relay cropping. In context of an orchard or a plantation of perennial fruit trees, however, the practices of growing annuals or relatively short duration crop in the interspace during their formative years is referred to as intercropping and the growing of perennial in the interspacing of perennials is called mixed cropping. The term multi-storey croppingrefers to a multispecies crop combination involving both annuals and perennials with an existing stand of perennials.

Intercrops with different fruitcrops under rained condition

S.No.	Crop	Duration for intercrop	Recommended Intercrops in rainy season
1	Kinnow	1-5 years	Onion, Black gram, Turmeric, Brinjal, Chillies and Okra Tomato, Cabbage, Pea Beans, Early potato and Cucurbits,
2	Aonla	1-5 months	Green gram, Cowpea, Cauliflower, Cabbage, Yam, Elephant foot,
3	Ber	1-4 years	Green gram, Moth, Cluster bean, Cowpea, Cumin and Chillies
4	Citrus	years	Beans, Carrots, Tomatoes, Berseem, Senji, Onion, Potato, Chillies, Pulses, Cucurbits, Okra, Gram, Peas, Potato and Cabbage
5	Guava	1.5 years	Onion, Black gram, Turmeric, Brinjal, Chillies and Okra Tomato, Cabbage, Pea Beans, Early potato and Cucurbits
6	Phalsa	2 year	Beans, Carrots, Tomatoes, Berseem, Senji, Onion, Potato, Chillies, Pulses, Cucurbits, Okra, Gram, Peas, Potato and Cabbage
7	Karonda	2 years	Beans, Carrots, Tomatoes, Berseem, Senji, Onion, Potato, Chillies, Pulses, Cucurbits, Okra, Gram, Peas, Potato and Cabbage

8

Training and Pruning

Importance of training and pruning

i. To control the habit of vegetative growth and regular bearing.

ii. Remove all limbs and dead branches.

iii. To improve flowering, fruiting and fruit quality

iv. Remove all broken shoots, diseased, or pest infection branches from all trees and shrubs.

v. It is usually best to prune temperate fruits and shrubs during early spring before full leaf.

vi. Maintain balance between shoots and roots at planting and nutrients uptake

vii. Provide the basic tree form and to aid the development of a strong tree framework by encouraging strong wide crotch angles

viii. Stimulate shoot growth near a cut.

ix. Permit proper light distribution and penetration to the tree canopy.

x. Increase insect and disease control by allowing for better spray penetration.

xi. Facilitate harvesting.

xii. Achieve orchard uniformity.

xiii. Produce the proper amount of well distributed fruiting wood.

xiv. Encourage a structurally sound tree by removing branches with weak narrow crotch angles or those with diseased, damaged, or unproductive wood. Invigorate and renew the bearing area of older trees.

Training

Physical techniques that control the shape, size and direction of plant growth are known as 'training' or improve appearance and horizontal orientation of branches make them fruiting better.

Objectives

- To develop of a strong tree framework by encouraging strong wide crotch angles and speedy winds.
- To allow uniform light distribution and penetration to the tree interior.

Details of Training

1. **Height of the head**: This is the height from ground to first branching or scaffolding. Depending on the height the trees could be divided in three groups.
    a) **Low head**: 0.6—0.9 m. This is common in windy areas. Such plants are easy to maintain.
    b) **Medium head**: 0.9—1.3 m. This is the most common height which combines both effects, ability to stand against wind and easy management.
    c) **High head:** More than 1.3 m. Common in tropics in wind free areas. Operations under the canopy are easy to perform.
2. **Number of scaffold branches:** It refers to allowing of number of scaffolds on the primary axis of the tree which vary from 2 to 15 but extremes are undesirable. In fruit trees, 5 to 8 scaffolds are preferred to make the tree mechanically strong and open enough to facilitate cultural operations.
3. **Distribution of scaffolds**: Scaffolds should be distributed in all the directions spaced at 45-60 cm allowing strong crotches through wide angles of emergence.

Types of training system

The woody perennials, which are widely spaced and remain on a place for a long duration, are trained for develop strong framework for sustainable production of quality produce and for ornamental beauty in different shapes (topiary). In these plants following types of training are followed.

i. **Open centre system ('V' shaped)**: The open centre system the main stem is allowed to grow to a certain height and the leader is cut to encourage lateral scaffold from near the ground giving a vase shaped plant. This is common in peaches, apricots, ber and Aonla.

ii. **Central leader system (closed centre):** The central leader system the central axis of plant is allowed to grow unhindered permitting branches all around. This system is also known as closed centre system and common in use in mango and sapota.

iii. **Modified leader system:** This system is in between open centre and central leader system wherein central axis is allowed to grow unhindered upto 4-5 years and then the central stem is headed back and laterals are permitted. It is common in apple, pear, cherry, plum, guava.

iv. **Cordon system**: This is a system wherein espalier is allowed with the help of training on wires. This system is followed in vines incapable of standing on their stem. This can be trained in single cordon or double cordon and commonly followed in crops like grape and passion fruit.

Open centre system

Central leader system

Modified leader system

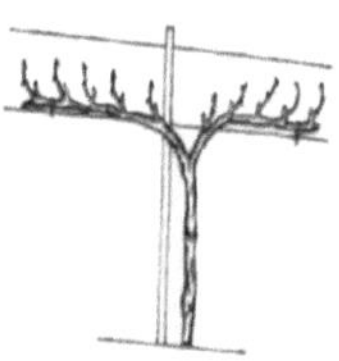
Cordon system

v. **Head System:** It is mostly used for spur bearing grape cultivars. In this system, vines are trained like a smallbush. Vines are allowed to, grow upto 1.2 meters, and then headed back to produce laterals. Four laterals- one in each direction is allowed to grow and rest are thinned out. In next dormant season, these laterals are cut back to 2 buds and further two arms of 20-30 cm are allowed on each secondary arm. After 3-4 years these vines will give a dwarf bush like appearance and requires no staking. Other training system which requires no staking are Palmette, Spindle bush, Dwarf pyramid and Head and spread systems.

vi. **Bower system:** It is also called as "Pandal" or "Arbour" or "Pergola" system. It is generally practiced in grapes and other cucurbitaceous vegetables like snake gourd, ribbed gourd, bitter gourd etc. In this system, the vines are spread over a criss cross net work of wires, usually at 2.1 to 2.4m above ground, supported by concrete or stone pillars. The vine is allowed to grow single shoot till it reaches the wire net and is usually supported by bamboo sticks tied with jute thread. When the vine reaches the wires, its growing point is pinched off to facilitate the production of side shoots.

vii. **Kniffin system:** In this system, two trellis of wire are strongly supported by vertical posts. The vines such as grape when trained in this system has four canes one along each wire and the bearing shoothangs freely with no tying being necessary.

Overhead trellis or Telephone system

This system consists of 3 or 4 wires usually kept at 45-60 cm apart fixed to the cross-anglearms supported by vertical pillars or posts. Vines are allowed to grow upto the height of 1.5 to 2.0m and then trained on this system. Moderately vigorous cultivars with apical dominance are best trained on such system.

Tatura trellis

In this system, trees are trained to a multi-layered wire trellis. The trellis is V-shaped, supported by two long, stout poles embedded into the soil angles of 60 cm from the horizontal. Five wires at 60 cm intervals are fastened to these poles. This system is being now followed for pome fruits, nut fruits and grapes. A well trained tree is an asset to the farmer and therefore, efforts should be made for training trees appropriately in formative years for sustainable production. In fact the process should have begun from nursery itself.

Pruning

Moderate pruning (removal of 25 to 30% of the canopy) is often done to reduce the canopy height or width of large mango trees. Moderate and even severe pruning does not cause injury to trees, but reduces production for one to several seasons. Pruning of fruit trees is usually carried out to shape and open up the canters, allowing free movement of air and sunlight into the tree. The ability of sunlight to penetrate the tree enhances the color of the fruit and improves quality. The ideal tree should have three and not more than four main trunks, be open inside, low-set, and 12 to 15 feet tall. Over this height harvesting becomes difficult. Trees must be pruned after harvesting. Pruning is one of the important operations in perennial fruit crops to harness the solar energy and utilize the available space, soil nutrient and moisture in newly establishing as well as existing productive and old senile fruit orchards. Understanding the problem with existing system of canopy management and measure to upon with respect knowledge is key to it. It has been observed that lack of focus upon regulating the growth and development of tree and appropriate canopy architecture right from the establishment stage is the main reason for such problems. As a result of non serious approach and negligence, the evergreen trees acquire their natural shape, which is very often not ideal for quality production. A majority of tree attains tall, upright and curved growth structure and the canopy is marked with cries- cross branches leading to a highly dens vegetative mass with very poor penetration of photosynthetic active radiation (PAR).

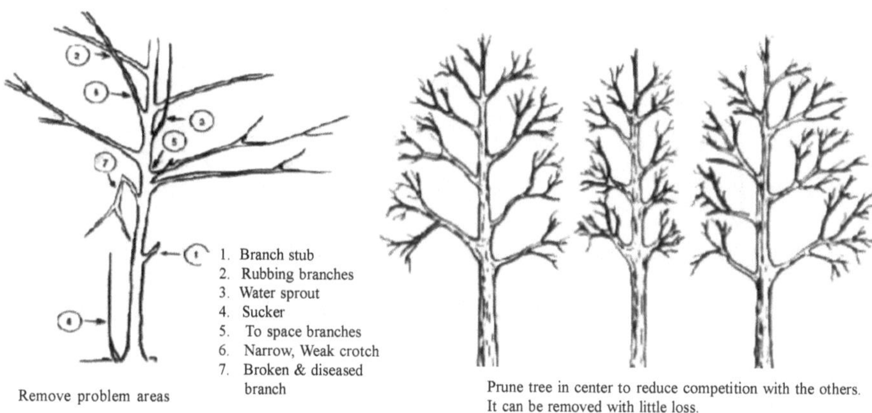

Remove problem areas

Prune tree in center to reduce competition with the others. It can be removed with little loss.

Such condition not only affect the photosynthetic rate but also with proliferation of pests and diseases, as they prefer shady condition. Consequently, with pressure the pest abundance and poor photo synthetically efficiency every the years in want of due care of tree canopy. The trees tune senile, unproductive and uneconomical. Therefore, may be explained as designing the plant as per need, using inherent plant characteristic in accordance with given set of condition and resources to perform the plant maximum. Proper design and shape influences light interception with assured higher monetary retunes to fruit growers. Early height control and tree canopy management are important techniques, which should be practiced in fruit crops for high returns. Beside internal factors, several other factors like planting system, spacing, and solar radiation. Wind velocity, rainfall pattern and light distribution etc., also play a very crucial role in the plant canopy development in fruit crops.

Make pruning cuts appropriately

Sharp pruning equipments are required to make the good structure. Do not leave stubs since they usually die back resulting in decay that can be serious, especially if large branches on the main trunk of the plant are involved. Once die back starts, the disease may spread easily to perfectly healthy tissue. The problem is the same if the branches are broken off rather than cut. Avoid tearing the bark when removing large branches. It is better to make a jump cut to relieve the weight of the branch; then make the cut flush with the limb or point of origin. Some specific rules are given in the discussion on how to prune various plants. Remember, no two plants are exactly the same so pruning techniques vary between plant species completely. The remaining branch assumes apical dominance over other cutoff branches.

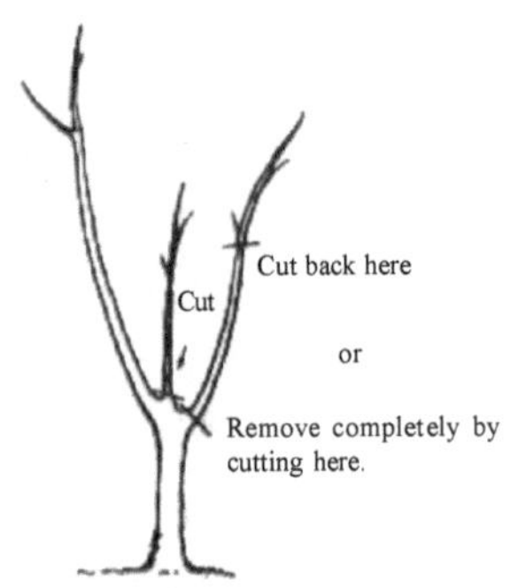

Leader has been choked out by rapidly growing laterals. The lateral shoots should have been pinched or removed much earlier, The correct the situation remove the original leader (middle) as well as the right lateral branch.

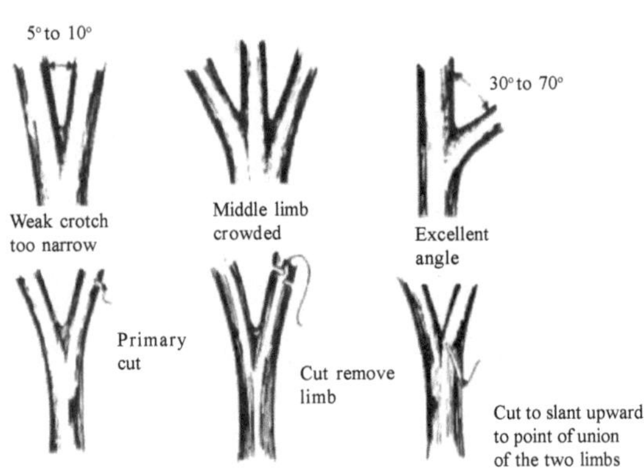

Pruning procedure for eliminating a narrow crotch

Most woody plants fall into two categories based on the arrangement of the buds on the twigs and branches. In general, the bud arrangements determine the plant's typical growth habit. Buds may have an alternate or an opposite arrangement on the twigs. A plant with alternate buds usually is rounded, pyramidal, inverted pyramidal or columnar in shape. Plants having opposite buds rarely assume any form other than that of a rounded tree or shrub with a rounded crown. The position of the last pair of buds always determines the direction in which the new shoot will grow. Buds on top of the twig probably will grow upward at an angle and to the side on which it is directed. In most instances, it is advisable to cut back each stem to a bud or branch. Selected buds that point to the outside of the plant are more desirable than buds pointing to the inside. By cutting to an outside bud, the new shoots will not grow through the interior of the plant or crisscross. To open up a woody plant, prune out some of the center growth and cut back terminals to the buds that point outward. In shortening a branch or twig, cut it back to a side branch and make the cut ½ inch above the bud. If the cut is too close to the bud, the bud usually dies. If the cut is too far from the bud, the wood above the bud usually dies, causing dead tips on the end of the branches. When the pruning cut is made, the bud or buds nearest to the cut are usually the new growing point. When a terminal is removed, the nearest side buds grow much more than they normally would since apical dominance has been removed, and the bud nearest the pruning cut becomes the new terminal. If more side branches are desired, remove the tips.

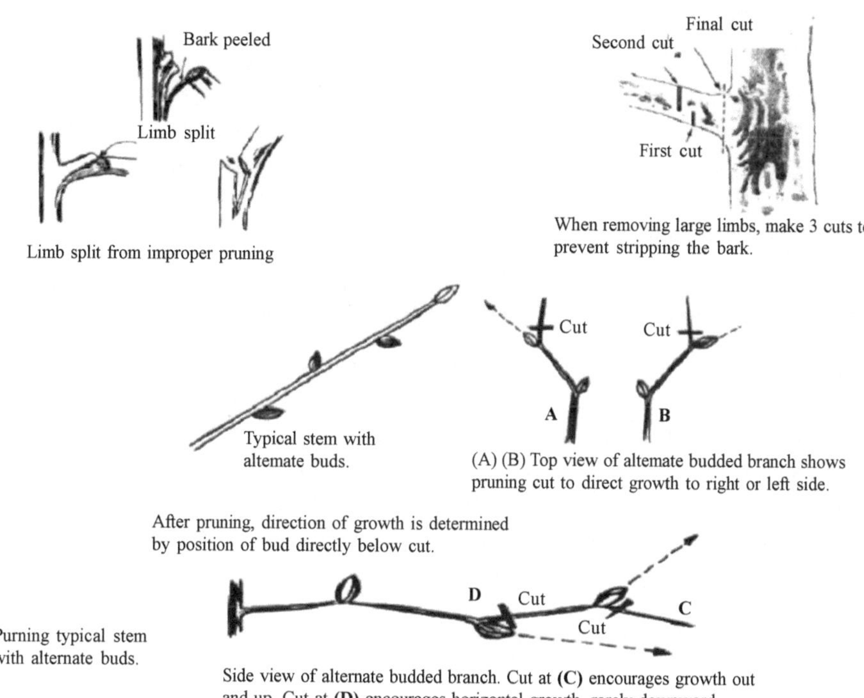

Limb split from improper pruning

When removing large limbs, make 3 cuts to prevent stripping the bark.

Typical stem with altemate buds.

(A) (B) Top view of altemate budded branch shows pruning cut to direct growth to right or left side.

After pruning, direction of growth is determined by position of bud directly below cut.

Purning typical stem with alternate buds.

Side view of alternate budded branch. Cut at **(C)** encourages growth out and up. Cut at **(D)** encourages horizontal growth, rarely downward.

The strength and vigour of the new shoot is often directly proportional to the amount that the stem is pruned back since the roots are not reduced. For example, if the deciduous shrub is pruned to 1 foot from the ground, the new growth will be vigorous with few if any flowers the first year. However, if only the tips of the old growth are removed, most of the previous branches are still there and new growth is shorter and weaker. Flowers are more plentiful although smaller. Thus, if a larger number of small flowers and fruits are desired, prune lightly. If fewer, but high quality blooms or fruits are wanted in succeeding years, prune extensively. Dehorning or cutting back mature trees and leaving large stubs because weakened trees with short, life spans. This commonly is done on mimosa trees, and most people think this is an acceptable process. However, it shortens the life of the tree and weakens it so it loses its natural shape and resistance to insects and diseases. It is better to maintain the natural shape of the tree when pruning. The height of a tree or shrub with two or more stems of equal size and vigour competing for dominance can be controlled by the length they are cut back. A tree or shrub with two branches growing at the same height is commonly known as a split crotch, double leader or weak crotch. For a stronger plant that can take the ice and wind better, cut one branch back or remove it completely. The remaining branch assumes apical dominance over other cut-off branches.

The objectives of pruning are

1. To remove dead or diseased wood
2. To remove additional growth flushes to allow more light penetration into the leaf canopy
3. To control tree height to facilitate cultural management practices.

Purposes of Pruning

1. Branches elongate from buds.
2. Branches increase in diameter from the cambium.

Mineral nutrients and water travel up from the roots system through the wood or xylem into the leaves. Here, in the leaves, food is manufactured and sent back through the phloem out to feed all parts of the plant, twigs, buds, flowers, roots, etc. If the terminal buds are removed, or twig end cut off side branching is induced, and a more compact habit of growth is obtained. If side branches or laterals are removed, a more upright on sider rejuvenation. Remove at least half of the existing old canes at ground level.

Crossed branches are unhealthy

When branches cross, there is a likelihood of the branches rubbing or growing together and creating a weak point in the tree as well as an entry point for pathogens. The smaller or weaker of the crossing branches should be removed.

Dead wood

It is offers a prime entry point for wood borers, as well as pathogens that can damage your tree. Carefully cut away any dead wood from the tree while avoiding cutting into the healthy portion of the tree. When dead wood is left in a tree, it is a hazard as it may fall at any time.

Thinning cuts

Thinning cuts are made on a branch all the way back to a major limb or trunk of the tree. They generally remove a dead branches.

Heading cuts

Heading cuts (tipping) are made toward the end of a branch near a node or growing point. Heading cuts are used to make a branch break buds and create multiple new shoots (or limbs). This is a common pruning cut used to train young trees.

Dormant pruning

Dormant pruning as new growth comes up - keep terminal growth pinched back to induce side branching and compact growth.

9

Research Organisations in Horticulture

1. Indian Institute of Horticultural Research (IIHR), Bangalore
2. Indian Institute of Vegetable Research (IIVR), Varanasi
3. Indian Institute of Spices Research (IISR), Calicut, Kerala
4. Central Institute of sub-tropical Horticulture (CISH), Lucknow
5. Central Institute of Temperate Horticulture (CITH), Srinagar
6. Central Potato Research Institute (CPRI), Kufri, Shimla
7. Central Tuber Crops Research Institute (CTCRI), Thiruvananthapuram, Kerala
8. Central Plantation Crops Research Institute (CPCRI) Kasargod, Kerala
9. Central Institute of Arid Horticulture (CIAH), Bikaner, Rajasthan
10. Central Institute of Post Harvest Engineering and Technology (CIPHET), Ferozepur, Punjab
11. ICAR Research Complex for Goa, Ela, Old Goa
12. ICAR Research Complex for North Eastern Hill Region. Barapani, Meghalaya
13. National Research Centre for Banana. Trichirapalli, Tamil Nadu
14. National Research Centre for Citrus, Nagpur, Maharastra
15. National Research Centre for Onion and Garlic, Pune, Maharastra
16. National Research Centre for Grape, Pune, Maharastra
17. National Research Centre for Medicinal and Aromatic Plants, Anand, Gujarat
18. National Research Centre for Mushroom, Solan
19. National Research Centre for Orchid, Gangtok, Sikkim
20. National Research Centre for Cashew nut, Puttur, Karnataka
21. National Research Centre for Seed Spices, Ajmer, Rajasthan
22. National Research Centre for Oil Palm, Eluru, Andhra Pradesh
23. National Research Centre for Pomegranate, Solapur, Maharastra
24. National Research Centre for Makhana, Patna, Bihar
25. National Research Centre for Litchi, Muzaffarpur, Bihar
26. National Horticulture Board (NHB), Gurgaon, Haryana

Post Harvest Technology

Name of institute	Location
Bhabha Atomic Research Centre (BARC)	Mumbai, Maharashtra
Central Research Institute for Post Harvest Engineering and Technology (CIPHET)	Ludhiana, Punjab
Central Food Laboratory (CFL)	Kolkata, West Bengal
Central Food Technological Research Institute (CFTRI)	Mysore, Karnataka
Defence Food Research Laboratory (DFRL)	Mysore, Karnataka
Food Preservation and Canning Institute (FPCI)	Lucknow, Uttar Pradesh
Food Research and Standardization Technology (FRSL)	Ghaziabad, Uttar Pradesh
National Institute of Food Technology Entrepreneurship and Management (NIFTEM)	Kundli, Haryana
National Agricultural Cooperative Marketing Federation of India ltd. (NAFED)	New Delhi
Public Health Laboratory (PHL)	Pune, Maharashtra
Indian Institute of Integrative Medicines (IIIM) earlier RRL	Jammu
Agricultural and Processed Food Products Export Development Authority (APEDA)	New Delhi
Indian Institute of Crop Processing Technology (IICPT)	Thanjavur, Tamil Nadu

Other institutes

Name of institute	Location
Central Plant Protection Training Institute (CPPTI)	Hyderabad, Andhra Pradesh
National Research Centre for Integrated Pest Management (NRCIPM)	New Delhi
National Institute of Hydrology (NIH)	Roorkee, Uttarakhand
National Bureau of Plant Genetic Resources (NBPGR)	New Delhi
National Bureau of Agriculturally Important Insects (NBAII)	Bengaluru, Karnataka
National Bureau of Agriculturally Important Micro-organisms (NBAIM)	Kushmaur, Uttar Pradesh
National Research Centre on DNA Fingerprinting	NBPGR, New Delhi
CSIR Laboratory	Palampur, Himachal Pradesh

National Centre for Integrated Pest Management (NCIPM)	New Delhi
Central Plant Protection Training Institute (CPPTI)	Hyderabad
National Horticulture Board	Gurgaon, Haryana
Indian Institute of Biodiversity	Itanagar, Arunachal Pradesh
Institute of Bioresource Management and Sustainable Use	Manipur
Central Fertilizer Quality Control Institute and Training Institute	Faridabad, Haryana
Indian Institute of Agricultural Biotechnology	Ranchi, Jharkhand
National Institute of Biotic Stress Management	Raipur, Chhattisgarh
National Institute of Abiotic Stress Management	Malegaon, Maharashtra

Global Horticulture Institutes

Name of the institute	Location
World Vegetable Centre (previously Asian Vegetable Research and Development Centre, AVRDC)	Taiwan
Association of Natural Rubber Producing Countries (ANRPC)	Kuala Lumpur, Malaysia
Biodiversity International (previously IPGRI)	Rome, Italy
International Food Policy Research Institute (IFPRI)	Washington, USA
International Fruit Genetics (IFS)	California, USA
Global Horticulture Initiative (GHI)	Rome, Italy
John Innes Horticultural Institute (JIHI)	England
Horticulture Research International (HRI)	Wellesbourne, UK
International American Spice Trade Associations	Washington D.C., USA
International Society for Horticultural Science (ISHS)	Belgium
International Institute of Horticulture (IIH)	Brazil
International Crop Research Institute for Semi-arid Tropics (ICRISAT)	Hyderabad, Andhra Pradesh
India Meteorological Department (IMD)	Pune, Maharashtra
International Gene Bank of Coconut for South Asia	Kidu, Karnataka
International Coconut Genetics Network (ICGH)	Rome
International Coconut Gene Bank (ICGB)	Indonesia, Brazil, Papua New Guinea and Cote d Ivoire

International Coconut Gene Bank- South Asia (ICG)	Kidu, Karnataka
Asia Pacific Coconut Community (APCC)	Vietnam
International Cocoa Gene Bank (ICG)	Trinidad and CEPLAC, Brazil
International Pepper Community (IPC)	Jakarta, Indonesia
International Network for Improvement of Banana and Plantain (INIBAP)	Montpellier, France
International Flower Market (IFM)	Alsmeer, Netherland
International Potato Centre (CIP)	Lima, Peru
International Rubber Research and Development Board (IRRDB)	Kuala Lumpur, Malaysia
International Rubber Association	London
World Coconut Germplasm Centre (WCG)	Andaman and Nicobar Islands
Institute of Enology and Viticulture	Japan
Institute of Grapevine Breeding	Germany
National Wine and Grape Research Centre	Australia
French National Institute for Agricultural Research (INRA)	France
International Seed Testing Association (ISTA)	Zurich, Switzerland
National Centre for Medium Range Weather Forecasting (NCMRWF)	New Delhi
World Cocoa Foundation	USA
International Centre for Underutilized Crops (ICUC)	Sri Lanka
Taiwan Banana Research Institute (TBRI)	Taiwan
International Institute of Tropical Agriculture (IITA)	Nigeria
Honduran Agricultural Research Foundation (FHIA)	Honduras
United Nations Environment Programme (UNEP)	Nairobi
International Organisation for Standardisation (ISO)	Budapest, Hungary
International cut flower growers association	USA
Association of cut Flower Growers	Haslett, USA
Association of Specialty Cut Flower Growers	West College Oberlin, USA
Asia and Pacific Seed Association	Bangkok, Thailand
International Protea Association	South Africa
International Flower Bulb Centre	Poland
International Bulb Research Centre	Vennestraat, the Netherlands
International Fragrance Research Association (IFRA)	Switzerland